Corinna Schnöller

A parasite immunomodulator and infection in models of allergic disease

Corinna Schnöller

A parasite immunomodulator and infection in models of allergic disease

Influence of a nematode immunomodulator and a gastrointestinal nematode infection on two models of allergic disease

Südwestdeutscher Verlag für Hochschulschriften

Impressum/Imprint (nur für Deutschland/only for Germany)
Bibliografische Information der Deutschen Nationalbibliothek: Die Deutsche Nationalbibliothek verzeichnet diese Publikation in der Deutschen Nationalbibliografie; detaillierte bibliografische Daten sind im Internet über http://dnb.d-nb.de abrufbar.

Alle in diesem Buch genannten Marken und Produktnamen unterliegen warenzeichen-, marken- oder patentrechtlichem Schutz bzw. sind Warenzeichen oder eingetragene Warenzeichen der jeweiligen Inhaber. Die Wiedergabe von Marken, Produktnamen, Gebrauchsnamen, Handelsnamen, Warenbezeichnungen u.s.w. in diesem Werk berechtigt auch ohne besondere Kennzeichnung nicht zu der Annahme, dass solche Namen im Sinne der Warenzeichen- und Markenschutzgesetzgebung als frei zu betrachten wären und daher von jedermann benutzt werden dürften.

Verlag: Südwestdeutscher Verlag für Hochschulschriften GmbH & Co. KG
Dudweiler Landstr. 99, 66123 Saarbrücken, Deutschland
Telefon +49 681 37 20 271-1, Telefax +49 681 37 20 271-0
Email: info@svh-verlag.de

Approved by: Berlin, Humboldt Universität zu Berlin, Diss., 2009

Herstellung in Deutschland:
Schaltungsdienst Lange o.H.G., Berlin
Books on Demand GmbH, Norderstedt
Reha GmbH, Saarbrücken
Amazon Distribution GmbH, Leipzig
ISBN: 978-3-8381-2780-4

Imprint (only for USA, GB)
Bibliographic information published by the Deutsche Nationalbibliothek: The Deutsche Nationalbibliothek lists this publication in the Deutsche Nationalbibliografie; detailed bibliographic data are available in the Internet at http://dnb.d-nb.de.

Any brand names and product names mentioned in this book are subject to trademark, brand or patent protection and are trademarks or registered trademarks of their respective holders. The use of brand names, product names, common names, trade names, product descriptions etc. even without a particular marking in this works is in no way to be construed to mean that such names may be regarded as unrestricted in respect of trademark and brand protection legislation and could thus be used by anyone.

Publisher: Südwestdeutscher Verlag für Hochschulschriften GmbH & Co. KG
Dudweiler Landstr. 99, 66123 Saarbrücken, Germany
Phone +49 681 37 20 271-1, Fax +49 681 37 20 271-0
Email: info@svh-verlag.de

Printed in the U.S.A.
Printed in the U.K. by (see last page)
ISBN: 978-3-8381-2780-4

Copyright © 2011 by the author and Südwestdeutscher Verlag für Hochschulschriften GmbH & Co. KG and licensors
All rights reserved. Saarbrücken 2011

Table of Contents

1. Zusammenfassung 5
   Summary 7
2. Introduction 9
   2.1. Immune regulation by parasites 9
      2.1.1 Immune reaction to and immune evasion by parasites 9
      2.1.2. Influence of parasites on systemic immune responses (spillover mechanisms) 11
   2.2. Hygiene hypothesis 12
   2.3. Worms and allergy 13
   2.4. The filarial cystatin Av17 15
   2.5. Allergic disorders 18
3. Results 20
   3.1 Av17 and Allergic Airway Hyperreactivity (AHR) – a murine asthma model to reveal in vivo effects 20
      3.1.1. Application of filarial cystatin blocks allergic airway disease in a preventive and a pre-challenge approach 20
      3.1.2. Filarial cystatin inhibits the development of allergic airway hyperresponsiveness in preventive and pre-challenge setting 21
      3.1.4. Filarial cystatin targets macrophages 26
      3.1.5. Cystatin increases the number of regulatory T cells 29
      3.1.6. Inhibition of allergic responses by cystatin is dependent on IL-10 31
      3.1.7. Macrophages are main producers of Av17-specific IL-10 33
      3.1.8. Features of Av17 application in vivo outside inflammation settings 33
   3.2. Av17 in allergic skin-disease: ovalbumin-induced atopic dermatitis 36
      3.2.1. Treatment with filarial cystatin prevents eczema in murine atopic dermatitis 36
      3.2.2. Cell infiltration into challenged skin is altered by cystatin treatment 37
      3.2.3. Changes in OVA-specific IgE in sera upon cystatin-treatment 38
      3.2.4. Cytokine pattern in spleen indicates altered response to the allergen 39
      3.2.5. Local cytokine production in skin is altered by cystatin treatment 39
      3.2.6. Cystatin restores Treg numbers in mesenteric lymph nodes but does not alter inguinal lymph nodes 40
   3.3. Heligmosomoides polygyrus infection and allergy 41
      3.3.1. Infection with the gastrointestinal nematode *H. polygyrus* ameliorates allergic airway inflammation but not the development of atopic dermatitis 41
      3.3.2. Allergen-specific humoral immune response of worm-infected mice in the asthma and dermatitis model is altered 43
      3.3.3. Local and systemic cytokine analysis in asthma and dermatitis differ 44
      3.3.4. Cellular infiltration in atopic skin is partly altered by *H.polygyrus* infection 46
      3.3.5. *H. polygyrus* infection leads to induction of regulatory T cells in mesenteric and peribronchial but not in inguinal lymph nodes 47
4. Discussion 50
   4.1. Av17 and Allergic Airway Hyperreactivity (AHR) 50
      4.1.1. Cystatin treatment modulates the allergic Th2 response 50
      4.1.2. Role of the anti-inflammatory cytokines IL-10 and TGF-ß 52
      4.1.3. Involvement of macrophages 54
      4.1.4. Role of regulatory T cells (Tregs) 59
      4.1.5. Potential role of B cells 61
      4.1.6. Possible receptors translating Av17 signals 62

4.1.7. Proteinase inhibitor function of Av17 and role of DCs 63
4.1.8. Mechanisms of immunomodulation by other helminth components 64
4.2. Influence of filarial cystatin on atopic dermatitis 66
4.3. Worm infection and allergy: Comparing the influence of *H.polygyrus* in an asthma and a dermatitis model 70
5. Outlook 77
6. Material and Methods 78
  6.1. Molecular biology and protein biochemistry methods 78
    6.1.1. Expression and purification of recombinant *Acanthocheilonema viteae* cystatin 78
    6.1.2. Purification of recombinant Cysele2, DHFR and mAv17 78
    6.1.3. Endotrap system to remove endotoxin contaminations 79
    6.1.4. Limulus amoebocyte test (endotoxin measurement) 79
    6.1.5. Quantification of protein 79
    6.1.6. SDS-PAGE (sodium-dodecyl-sulfate polyacrylamid-gelelectrophoresis) and coomassie staining 79
    6.1.7. Dialysis 79
    6.1.8. RNA and cDNA preparation of lung and skin tissue and cultured cells 79
    6.1.9. Real time PCR (TaqMan-System) 80
  6.2. Cell culture techniques 81
    6.2.1. RBL assay 81
    6.2.2. Preparation of mesenteric lymph node cells (MLNC), splenocytes, inguinal (ILNC) and peribrochial lymph node cells (PBLNC) 81
    6.2.3. Peritoneal lavage and preparation of peritoneal exudate cells (PEC) 81
  6.3. Animal models 82
    6.3.1. Animals used for experiments 82
    6.3.2. Murine asthma model and measurement of airway hyperreactivity (AHR) .. 82
      6.3.2.1. Bronchoalveolar lavage (BAL) 83
      6.3.2.2. Histological analysis of lung 83
    6.3.3. Murine dermatitis model (atopic eczema) 84
      6.3.3.1. Histological analysis of skin 84
      6.3.3.2. Immunohistochemistry of skin sections 84
    6.3.4. Depletion of Treg cells and blocking of IL-10 receptor in vivo 85
    6.3.5. Depletion of macrophages / Preparation of multilamellar vesicles (MLV) containing clodronate 85
  6.4. Immunological methods 86
    6.4.1. Enzyme-linked immunosorbent assay 86
      6.4.1.1. Serum levels of IgE (total and OVA-IgE), OVA-IgG1, OVA-IgG2a 86
      6.4.1.2. Biotinylation of ovalbumin 86
      6.4.1.3. Subclass - specific ELISA 86
    6.4.2. Cytokine detection 87
    6.4.3. Flow cytometric analysis 87
    6.4.4. Proliferation assay 87
  6.5. Parasitological methods 88
    6.5.1. Life cycle of *Heligmosomoides polygyrus* 88
    6.5.2. Fecal egg count 88
    6.5.3. Preparation of *H. polygyrus* adult worm antigen 88
  6.6. Statistical analysis 88
  6.7. Material 89
    6.7.1. Laboratory equipment 89
    6.7.2. Buffers and media 89

  6.7.3. Chemicals, biologicals and recombinant cytokines 91
  6.7.4. Commercial Kits 91
  6.7.5. Antibodies and secondary reagents 92
  6.7.6. Software 92
7. Abbreviations 93
8. References 96

1. Zusammenfassung

Parasitische Würmer wie Filarien können das Immunsystem ihres Wirtes effektiv modulieren und so über viele Jahre im Wirt persistieren. Diese Immunmodulation schützt nicht nur den Parasiten, sondern kann auch einen Schutz des Wirtes vor allergischen Reaktionen vermitteln. Es scheint, dass Wurminfizierte weniger Allergien bekommen. Parasiten sezernieren immunmodulatorische Moleküle die in engem Zusammenhang mit Immunevasion stehen und starke anti-allergische und anti-inflammatorische Eigenschaften haben könnten. In der vorliegenden Arbeit wurde Filariencystatin, ein sezernierter Proteinaseinhibitor des Fadenwurms *Acanthocheilonema viteae* in Allergiemodellen eingesetzt.

Filariencystatin supprimierte Th2-assoziierte Entzündungen und die daraus resultierende Atemwegserkrankung in einem murinen Ovalbumin-induzierten Asthmamodell. Die Behandlung der Tiere mit Cystatin während und ebenso nach Abschluss der Sensibilisierungsphase und vor der Atemwegsprovokation mit Allergen, verhinderte die Atemwegsentzündung und –hyperreaktivität, reduzierte allergen-spezifisches IgE und regelte die IL-4 Produktion herab. Diese Effekte von Cystatin beruhen wahrscheinlich auf der Induktion von IL-10 sezernierenden Makrophagen. Dafür spricht, dass sowohl die Depletion von Makrophagen durch Clodronat gefüllte Liposomen, als auch die Blockierung von IL-10 durch anti-IL-10-Rezeptor Antikörper den schützenden Effekt von Cystatin aufhoben und die allergische Entzündung wiederherstellten. Im Gegensatz dazu hatte die Entfernung von regulatorischen T-Zellen durch anti-CD25 Antikörper nur sehr begrenzte Effekte auf die Cystatinwirkung.

In einem murinen Allergiemodell der Haut (Ovalbumin-induzierte atopische Dermatitis) konnte durch Cystatinapplikation die Entzündung der Haut und die allergische Erkrankung ebenfalls supprimiert werden. Die Ekzembildung und das Einwandern von inflammatorischen Zellen in die Hautläsionen war ebenso behindert, wie die Bildung von allergenspezifischem IgE. Die Verbesserung der atopischen Dermatitis führte zu einem dramatischen Anstieg von TGF-ß in den Ekzemregionen.

Zusammenfassend belegen die vorliegenden Daten, dass ein einzelnes rekombinantes Helminthenprotein den anti-allergischen Effekt einer Gesamtwurminfektion nachahmen kann. Ein weiterer Aspekt dieser Arbeit war herauszufinden, ob die Immunmodulation von Helminthen an die Biologie bzw. die Lokalisation des Wurmes gebunden ist. In zwei verschiedenen Allergiemodellen, der mucosa-assoziierten allergischen Atemwegsentzündung und der nicht-mukosa-assoziierten, kutanen atopischen Dermatitis wurde der Einfluss einer gastrointestinalen Wurminfektion mit *Heligmosomoides polygyrus* untersucht. Im

Asthmamodell konnte eine deutliche Reduzierung der Inflammation in wurminfizierten Mäusen beobachtet werden. Im Gegensatz dazu zeigten wurminfizierte Mäuse im Dermatitis Modell keinerlei Verbesserung des klinischen Bildes. Im Asthmamodell führte die Wurminfektion zu signifikant erhöhten Werten von Foxp3$^+$ regulatorischen T-Zellen (Treg) in peribronchialen Lymphknoten. Im Dermatitismodell hingegen konnten keine Treg Zellen in ekzematöser Haut nachgewiesen werden und auch in den drainierenden inguinalen Lymphknoten war kein Anstieg der regulatorischen Zellen zu beobachten. Das deutet daraufhin, dass durch eine gastrointestinale Nematodeninfektion induzierte Treg Zellen zwar in andere mukosale Gewebe wie die Lunge migrieren können, jedoch nicht in der Lage sind in nicht-mukosale, kutane Gewebe zu wandern. Diese Ergebnisse sind ein wichtiger Aspekt beim Einsatz von Helminthen zur Behandlung allergischer Erkrankungen, da die Lokalisation des Wurmes ein entscheidender Faktor in Richtung Verbesserung oder Verschlimmerung allergischer Erkrankungen zu sein scheint.

Summary

Influence of a nematode immunomodulator and a nematode infection on two models of allergic disease

Parasitic worms, like filariae, have the intriguing capacity to modulate immune responses directed against them, and therefore can persist up to years in a host. This immunomodulation does not only protect the parasite but was found to mediate a negative correlation between infections with parasitic worms and the prevalence of allergic disease. Secreted molecules of the parasites were found to be strongly involved in immune deviation and might bear themselves strong anti-allergic and anti-inflammatory potential. In the present study filarial cystatin, a secreted protease inhibitor of the filarial nematode *Acanthocheilonema viteae*, was found to suppress Th2-related inflammation and the ensuing asthmatic disease in a murine model of ovalbumin-induced allergic airway responsiveness. Treatment with recombinant filarial cystatin during or after sensitization and before challenging mice with the allergen, inhibited airway inflammation and airway hyperreactivity, reduced levels of allergen-specific IgE and down-regulated IL-4 production. These effects of Av17 on allergic airway inflammation were probably due to induction of IL-10 secreting macrophages. Depletion of macrophages by liposomes containing clodronate as well as blocking of IL-10 by application of anti-IL-10 receptor antibodies, prevented the curative effects and restored allergic airway inflammation. In contrast, depletion of regulatory T cells by anti-CD25 antibodies had only limited effects.

Treatment with cystatin in a model of allergic skin disease, namely ovalbumin-induced dermatitis, was determined to have suppressive activity on inflammation and the ensuing allergic disease, too. Eczema formation and recruitment of inflammatory cells to lesions was clearly impaired and also allergen-specific IgE production substantially reduced. The protective effect on atopic dermatitis was accompanied by a dramatic increase of TGF-β in the eczema region. Hence, the data demonstrate that treatment with a single recombinant helminth protein can exert the anti-allergic effects of helminth infections.

Another part of this study was to investigate if immunomodulation by helminths is dependent on the biology (localisation) of the worm. The influence of a gastrointestinal nematode infection with *Heligmosomoides polygyrus*, was analyzed in two distinct murine models of allergy: mucosa-associated airway disease and non-mucosal, cutaneous atopic dermatitis. Mice concomitantly infected with *H. polygyrus* clearly showed reduced signs of allergic airway inflammation. In contrast, no significant differences in phenotypical signs of dermatitis, such as severity of eczematous skin lesions, were observed between infected and control animals in the atopic dermatitis model. Worm infection was associated with significantly elevated numbers of Foxp3$^+$ regulatory T cells (Treg) in peribronchial lymph node cells in *H. polygyrus*-infected sensitized and airway challenged mice, whereas Treg cells were basically absent in eczematous skin of mice with experimental dermatitis and not up-regulated in their skin-draining lymph node cells. Therefore, it seems that Treg cells induced by a gastrointestinal nematode infection can migrate to mucosa associated lung lymph nodes, but are not able to migrate into cutaneous tissue. These findings might be an important aspect for future considerations of helminths for treatment of allergic diseases, as localisation of the parasites might be a crucial factor leading to amelioration or aggravation of allergic disease.

2. Introduction

2.1. Immune regulation by parasites

2.1.1 Immune reaction to and immune evasion by parasites

Parasites are defined as organisms that live in or on and obtain nourishment from another organism of a different species (Lucius 2008). Among multicellular parasites the parasitic nematodes have a well-balanced host-parasite-equilibrium and thereby display long persistence without gravely damaging their host. This equilibrium is partly due to the fact that helminths do not replicate in their hosts and high parasite loads can only be accounted for by continuous exposure and infection.

Hosts have a wide variety of mechanisms to fight parasites. As far as helminths are concerned, immune reactions comprise the whole repertoire of immune cells. A scenario might begin with antigen-presenting cells (macrophages, dendritic cells or B cells) that process the foreign worm antigens and display them to T-helper cells in a MHC class II-context. T-helper cells start to produce cytokines that activate and attract macrophages and other innate immune cells (eosinophils, neutrophils, basophils) as well as B cells and more T helper cells. Antigen-specific B cells differentiate to plasma cells and produce large amounts of antibodies. Antibodies can opsonise the pathogen to direct eosinophils or neutrophils, as well as macrophages, to their target (antibody-dependent cellular toxicity, ADCC). Also, antibodies bind to Fcɛ-receptors displayed by mast cells. This sensitizes mast cells, which then secrete large amounts of histamine and other mediators once antigen contact ensues, facilitating the attraction and accumulation of further immune cells. Furthermore, special populations of macrophages (alternatively activated macrophages generated in the presence of helminths) were shown to be involved in the termination of secondary infections (Anthony 2006).

Intriguingly, some parasitic filariae (*Onchocerca volvolus*) can persist in their hosts for up to 15 years (Plaisier 1991). This means that these creatures must have developed strategies to escape the host's immune response in order to survive in the long run. These so-called immune evasion mechanisms comprise a large repertoire and lead to avoidance of nearly any immune response set up by the host.

Helminths protect themselves with manifold strategies: their surface, the multilayer cuticula, prevents quite a lot of immune reactions, therefore reacting as a biological armour. Furthermore, helminths migrate in their hosts and can evade pathology by moving away from inflammation. Another evasion mechanism exploited, for example by filariae, is their location in immunological privileged tissue, such as nodules. Additionally, some parasites secrete proteases to prevent antibody opsonization and they are able to produce anti-oxidative enzymes. Helminths can induce and suppress immune reactions to aid their survival, partly systemically and partly locally in direct interaction with immune cells (Maizels 2004).

Millions of people world-wide are infected with helminths. Generally, only a small percent of infected individuals suffer from severe symptoms, but most remain asymptomatic (van Riet 2007). Helminths are known to skew the immune response towards Th2, characterized by Th2 related cytokines IL-4, IL-5 and IL-13 that induce B lymphocytes to switch to IgE antibody production. However the role of Th2 has been described to be that of eradicating helminth infections (Finkelman 2004).

The question, therefore, is why helminths carry Th2 inducing molecules? Presumably these molecules are essential for parasite biology, as could be demonstrated for the filarial protein tropomyosin (Sereda 2008).

As mentioned before, helminths have developed different strategies for survival in their host. Schistosomes, for example, can compromise complement function (Ouaissi 1981), may degrade host immunoglobulins (Auriault 1981), can acquire surface molecules from their host to perform molecular mimicry and may produce cytokine mimics (Maizels 1993). Furthermore they can down-regulate B and T cell responses via induction of Tregs, anti-inflammatory cytokines or AAM (Maizels 2003).

Longevity of filariae (Subramanian 2004) reflects suppression or modulation of the host´s immune system (King 2001, Brattig 2004) such that there is an impairment of lymphocyte proliferation and bias in the production of cytokines and antibodies. Generally Th2 – associated IL-4 and anti-inflammatory IL-10 are increased and associated with elevated levels of IgG4 (mouse IgG1), an antibody subclass that is unable to activate complement or bind with high affinity to phagocytic cells, thereby having little value in the elimination of pathogens. Overall, filarial nematodes lead to a suppressed (impaired lymphocyte proliferation), anti-inflammatory (increased IL-10 and IgG4) and Th2-like (IL-4) immune response.

It is thought that such modulation does not only lead to parasite survival, but is conducive to host-health by limiting pathology resulting from aggressive immune responses (immunopathology) (Harnett 2006).

Immunomodulation is believed to be beneficial to both host and parasite, as it protects helminths from being eradicated, but at the same time safeguards the host from excessive immune responses that might lead to pathology. However, it does usually only occur in chronic or high level infections. Interestingly, modulation of immune response seems not only to be directed against the worms but also to non-related bystander antigens (van Riet 2007).

2.1.2. Influence of parasites on systemic immune responses (spillover mechanisms)

On the one hand, helminth modulation, as in infection with *Heligmosomoides polygyrus*, can impair host resistance to other pathogens like *Citrobacter redentium* and enhance Citrobacter-induced colitis (Chen 2005). On the other hand, worm infections (*Ascaris lumbricoides*) protect hosts from the negative effects of *Plasmodium falciparum* (cerebral malaria and acute renal failure) (Nacher 2000, 2001) and *H. polygyrus* protects from intestinal pathologies, e.g. of *Helicobacter pylori* (Fox 2000, Elliott 2004).

The immune modulation by helminths has been explained by induction of Treg cells, which inhibit the recruitment and the activation of effector cells (Wilson 2005, Baumgart 2006, Metwali 2006), by driving the balance between different antibody isotypes (Mutapi 2005) and/or T-cell populations (McKee 2004, Gillian 2005). Among the potential mechanisms of helminth immunomodulation is also the concept of secreted/released immunomodulators. Up to now, cytokine homologs, protease inhibitors, lectins and glycans that may misdirect or interrupt host immune response were described.

Homologs of mammalian cytokines were found in *Brugia malayi* and *pahangi* (Gomez-Escobar 1998) and in *Schistosoma japonicum* (Hirata 2005) and described to be parasite-encoded TGF-ß homologs. Protease inhibitors were described as molecules that interfere with antigen-processing and presentation of antigen. One of the most important types are cystatins (Hartmann 2003; see section 2.1.3.).

Several glycans derived from Schistosoma eggs were found to expand a $Gr1^+$ suppressor macrophage population in the mouse peritoneal cavity that produces high levels of TGF-ß and IL-10, thereby suppressing proliferation of naïve T cells *ex vivo* (Terrazas 2001).

In addition *N. brasiliensis* has been described to release a component with acetylhydrolase activity that leads to cleavage of platelet-activating factor (PAF). PAF has been shown to promote bronchoconstriction, increase vascular permeability and activate inflammatory leukocytes (Blackburn 1992). Excretory/secretory (E/S) products of the hookworm *Necator americanus* are able to cleave eotaxin, thereby inhibiting the recruitment and activation of eosinophils (Culley 2000).

Interestingly helminths can, but usually don´t induce strong allergic responses in humans (Cooper 2002). There are close similarities between allergic inflammation to environmental allergens and to parasite antigens. Both are associated with high levels of IgE, tissue eosinophilia and mastocytosis, mucus hypersecretion and T cells preferentially secreting type 2 cytokines as IL-4, IL-5 and IL-13 (Yazdanbakhsh 2002, Cooper 2004). In human studies it turned out that acute infections may lead to allergic syndromes as Loeffler's syndrome, which occurs when *A. lumbricoides* larvae migrate through the lungs for the first time (Cooper 2002). However, in individuals with chronic infections clinical allergic reactions appear to be rare. Most intriguingly, Schistosomiasis may decrease immune response to allergens and clinical manifestation of asthma (Araujo 2006).

Thus, helminth infections – although responsible for an overall negative impact on the host's health – bear the intriguing potential to positively modulate pathological inflammatory responses. Evidence for this hypothesis was provided by clinical trials in patients with inflammatory bowel disease, in which the eggs of intestinal worms reduced symptom severity (Summers 2005 a, b).

2.2. Hygiene hypothesis

ISAAC (International study of asthma and allergies in childhood) and ECRHS (European community respiratory health survey) studies show that asthma prevalence has increased steadily over the last decades (Pearce 2000).

In 1989 Strachan formulated the hygiene hypothesis in which he suggested that the increasing incidence of asthma and allergies could be attributed to reduced exposure to childhood Th1 polarising infections, because of an increase in vaccination and improvements in sanitation. Lacking an imprint of Th1-biased memory cells, the inherent Th2 bias of immune responsiveness at mucosal surfaces would proceed unchecked and lead to allergies in genetically predisposed individuals. Today, the hypothesis has been modified mainly concentrating on exposure to endotoxin in childhood, thereby leading to a Th1 bias (Braun-Fahrlander 2002). However, an underlying universal immunological mechanism has still not been determined (Vercelli 2006).

Furthermore, there are several arguments against this Th1/Th2 counter-regulation model: both allergies and helminth infections are characterized by induction of Th2-polarized immune response and IgE, however in tropical countries with high prevalence of Th2-driving helminth infections, allergies are less common (ISAAC 1998). Moreover, the prevalence for autoimmune diseases like type 1 diabetes, multiple sclerosis or Crohn's disease (all with a Th1-biased inflammatory response) is increasing parallel to asthma incidences (Bach 2002).

2.3. Worms and allergy

A hypothesis, attempting to reconcile all these observations, is that the primary consequence of all pathogen infections is the induction of immunoregulatory mediators critical for the prevention of immune hyperreactivity (allergy and autoimmunity) (Wills-Karp 2001).

Over the last years there have been multiple reports about the correlation between worm infections and allergic disorders in human studies (Dagoye 2003, van den Biggelaar 2001, 2004, Araujo 2006). Studies on helminths revealed links between the suppression of allergic hyperreactivity and its reversal through treatment with anti-helminthics (Lynch 1993, van den Biggelaar 2004), and various immunological parameters to explain this phenomena were evaluated (e.g. for Schistosoma: van den Biggelaar 2000; Araujo 2004). Nevertheless, there are also reports indicating an enhancement of allergic inflammation in humans infected with *Ascaris lumbricoides* and *Anisakis simplex* (Kennedy 2000, Obihara 2006). More recent human studies indicate that worm infection has protective effects on asthma, but detrimental effects on dermatitis (Haileamlak 2005). Metaanalysis of recent publications were carried out by Leonardi-Bee and colleagues and they came to the conclusion that A*scaris* infections appear to increase incidence of asthma, whereas hookworm infestations seem to reduce incidences of asthma (Leonardi-Bee 2006).

As human studies have not always been consistent, a lot of studies have been done in animal models: amelioration of allergic asthma was found in infections with gut nematodes, male schistosomes and other helminths (Wilson 2005, Kitagaki 2006, Wohlleben 2004, Mangan, 2004, 2006) and effects on food allergy have also been reported (Nagler-Anderson 2006). A protective mechanism of *H. polygyrus* on allergic hyperreactivity has been shown to be dependent on $CD25^+CD4^+$ T cells in mesenteric (Wilson 2005) and peribronchial lymph nodes (Kitagaki 2006). Also, infection with *Strongyloides stercoralis* was found to suppress pulmonary allergic response in a preventive approach. Although not reducing eosinophilia in lungs, OVA-IgE levels were abolished and eotaxin levels strongly decreased (Wang 2001). Lately, infection with *Trichinella spiralis* was linked to inhibition of cellular recruitment to lungs of mice infected with *influenza A*, thereby reducing lung pathology, mostly due to TNF-α inhibition and not to expansion of IL-10 producing cell populations (Furze 2006). On the other hand, *Schistosomes* protect mice from anaphylaxis by IL-10-producing B cells (Mangan 2004) and protection of airway hyper-responsiveness by male schistosomes was correlated to increased IL-10 levels in lungs (Mangan 2006). *H. polygyrus'* dampening of food allergy caused by peanut is thought to be mediated via an IL-10-dependent regulatory mechanism (Bashir 2002). This protection could be transferred by $CD4^+$ T cells harvested from worm-infected MLN and spleens (Nagler-Anderson 2006). In addition, an influence on co-stimulatory molecules on APCs was reported (upregulation of CD80 and CD86), but only locally in the mucosa (Peyers patch and MLN) and not in the periphery (spleen). Moreover, suppression of allergen-induced airway inflammation by *Nippostrongylus brasiliensis* was not observed in IL-10-deficient mice (Wohlleben 2004). Within *Ascaris* infection, transfer of regulatory T cells of infected animals was capable of mediating amelioration of allergic eye disease (Schopf 2005). Somewhat conflicting results were observed when rats were infected with *Strongyloides venezuelensis,* concomitant with induction of airway inflammation. Negrao-Correa and co-workers reported prolonged pulmonary inflammation, but suppressed AHR during lung passage of the parasite, indicating a partial protective effect of worm infection (Negrao-Correa 2003).

Intensity and chronic nature of infection seem to be very important factors that are associated with amelioration of allergic inflammation. Whereas, in great contrast, acute infections with low worm burdens may procure devastation or aggravation of allergic responses (Smits 2007). It also seems to be crucial as to which parasite infects the patient/animal and which disease is monitored, as e.g. a study in Ethiopia revealed reduced wheeze with hookworm infection but increased allergen skin test reactivity with trichuriasis (Scrivener 2001).

In summary parasitic worm infections characteristically elicit Th2-polarized responses quite similar to atopic individuals. However, allergies seem to be less prevalent in helminth infected people. Parasites which can exploit host down-regulatory networks are likely to gain advantage in the battle for long-term survival in the host (Maizels 2003). The immune system may have evolved optimally in the regulated environment of infections, but now that the environment has become more hygienic sufferers might be prone to overzealous reactions to innocuous substances, generating the rapidly increasing levels of allergy and autoimmunity being experienced in the developed world (Wilson 2005).

Introduction

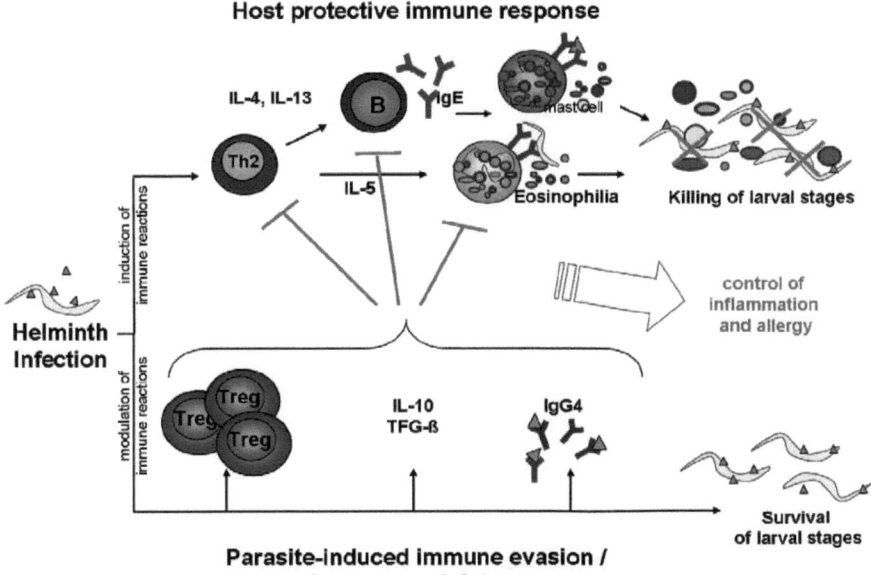

Figure 1: Helminth infections induce a Th2 response, partly because they harbour Th2 driving molecules (nematode allergen, small triangle). In a protective host response (upper panel) the cytokines interleukin-4 (IL-4), IL-5 and IL-13 are secreted, which lead to eosinophilia and production of allergy-associated IgE antibodies by B cells. Mediators released by activated eosinophils and mast cells kill opsonised larval stages. Successful helminths are able to evade the host's protective immune response (base panel). Parasite immunomodulators may induce regulatory T cells (Treg), immunosuppressive cytokines (IL-10, TGF-ß) and switching to competing IgG4 antibodies that do not activate eosinophils. Thereby killing of larvae is anticipated. Allergic immune responses resemble host protective immune responses regarding Th2 polarization, IgE, mast cells and eosinophilia. Blocking of these mechanisms does not only protect larvae but may also control inflammation and allergy.

Down-modulation of allergic inflammation has been explained as a survival strategy of worms. Well-adapted parasites can prevent the triggering of allergic effector mechanisms in order to modulate inflammatory host responses directed against them (Moncayo 2006). Figure 1 shows an overview of the interplay of host protective and parasite-induced evasion mechanisms.

As a side effect, worm-induced immune modulation can lead to down-regulation of other immune responses, such as allergic reactions in humans and experimental animals (Yazdanbakhsh 2002, Wilson 2005, Dunne 2005, Fallon 2007), and also of viral and bacterial infections (Fox 2000, Furze 2006).

2.4. The filarial cystatin Av17

Immunomodulatory effects of parasitic worms can be exerted by secreted compounds, which have supposedly been shaped and optimized during co-evolution with their vertebrate hosts.

Strong candidates for the immunomodulatoy components are parasite cystatins (Hartmann 2003, Maizels 2004).

Cystatin of the filaria *Acanthocheilonema viteae* is secreted by all developmental stages (Hartmann 1997) and can be found in excretory/secretory (E/S) products of worms in culture. As the larvae are found in blood and adult worms dwell in tissue the filarial parasite and the secreted cystatins can get in direct contact with the host's immune system.

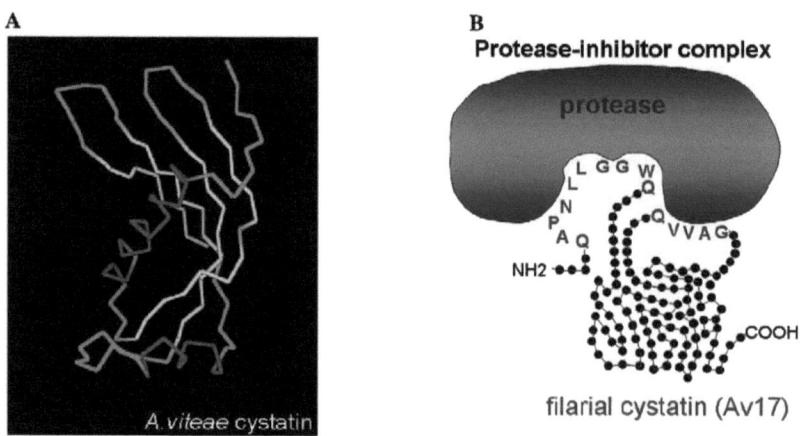

Figure 2: *Acanthocheilonema viteae* cystatin (Av17): (A) computer-generated model of *A. viteae* cystatin (generated by T. Buhrke) showing relevant region for its proteinase inhibitor function. (B) mechanistic model of interaction of Av17 with a protease and the protease-inhibitor complex. Letters indicate highly conserved regions binding into the active site of the protease.

Filarial cystatin belongs to the cystatin super-family that consists of three major families of evolutionary related, reversible tight-binding inhibitors of cysteine proteases (Nicklin 1984). Type I cystatins, the stefins, are unglycosylated proteins of about 11 kDa which lack a signal sequence and any disulphide bonds, and are predominately intracellular. Type II cystatins, the second family, are about 13–14 kDa in size, contain a signal sequence and two carboxy-terminal disulphide bonds and are usually exported from the cell. Type III cystatins, the kininogens, are glycosylated, multifunctional proteins with three cystatin domains that contain disulphide bonds and are synthesized in the liver and exported into the blood. They are the major inhibitors of cysteine proteinases there, along with alpha-2-macroglobulin (Barrett 1986, Abrahamson 1994).

Cystatins are found in all living organisms and regulate cysteine proteases which are involved in many biological processes, such as protein metabolism, antigen-processing and inflammation (Henskens 1996). Cysteine proteases are defined as enzymes that degrade polypeptides. Cysteine proteases have a common catalytic mechanism that involves nucleophilic cysteine thiol in a catalytic triade. Examples for cysteine proteases are papain, cathepsins, caspases and calpains. (en.wikipedia.org/wiki/Cysteine_protease). Cysteine proteases degrade proteins within the endolysosomal compartment of APCs. Furthermore, they are involved in the cleavage of the MHC class II-associated invariant chain, which leads to the formation of class II –associated invariant chain peptide (CLIP)-associated MHC molecules. By removing CLIP, the binding of peptides to MHC class II is possible (Watts 2001).

Acanthocheilonema viteae cystatin (model of the structure in Figure 2A) exhibits highest homology to the type 2 family cystatins. Its close relative, human cystatin C, is expressed and secreted constitutively by mononuclear phagocytes (Warfel 1987). Among others, the cystatins have important functions in the regulation of proteases relevant in immune responses. Thus, the endogenous inhibitor cystatin C influences the fate of newly synthesized peptide-MHC class II complexes by regulating the activity of cathepsin S, a cysteine protease that is essential in Ii-chain degradation in B cells and dendritic cells (Riese 1996, Pierre 1998, Nakagawa 1999). Additionally, human cystatin C has been described to inhibit the phagocytic function of monocytes and granulocytes (Leung-Tack 1990 a, b), and it has been shown that various members of the cystatin super-family up-regulate the inducible NO production of murine macrophages (Verdot 1999).

Schoenemeyer and colleagues were able to show that a cystatin secreted by the human filariae *Onchocerca volvolus* is also able to inhibit human cysteine proteases. The immunologically relevant cathepsins L and S were significantly down-modulated by *O. volvolus* cystatin (Schoenemeyer 2001) thereby altering antigen presentation of APCs (model of the interaction shown in Figure 2B). Furthermore *in vitro* studies on recombinant *O. volvolus* and *A. viteae* cystatin revealed that stimulation of human PBMCs or murine splenocytes led to secretion of high amounts of IL-10 and inhibition of T cell proliferation (Hartmann 2003). The effect was specific for recombinant cystatin of filarial nematodes, whereas recombinant cystatin of the free-living nematode *Caenorhabditis elegans* did not modulate macrophage functions *in vitro* (Schierack 2003). Cystatins of several species have been investigated (Onchocystatin in *Onchocerca volvolus*, Lustigmann 1992; Nippocystatin in *Nippostrongylus brasiliensis*, Dainichi 2001; *Litomosoides sigmodontis* cystatin, Pfaff 2002) and are seen to be associated with immunosuppression. The most information is known about Av17 and Ov17.

2.5. Allergic disorders

Allergic disorders like asthma, rhinitis and atopic dermatitis, affect 10 – 15 % of Western populations and their prevalence has doubled in the last 10-15 years (Kay 2001). Interestingly, in the past decades the frequency and severity of atopic disorders have steadily increased, particularly in developing countries (Braman 2006).

These diseases are caused by a dysregulated immune response and share a common etiology, in that the immune system is misguided and treats foreign antigens as allegedly harmful agents. Allergic disorders lead to production of IgE against harmless environmental antigens like pollen, animal proteins or house dust mites, and the induction of eosinophilia. A strong bias to a Th2-biased immune response, which is thought to be caused by a lack of danger signals mediating Th1 responses and uptake of allergens via mucosal surfaces, is thought to be the main reason for pathology (Umetsu 2006 a, b).

Asthma belongs to type I (immediate) hypersensitivity responses. Type I responses occur in two phases, the immediate reaction after a few minutes which is characterized by vasodilation, congestion and edema (mainly mediated by IgE and basophils) and the late phase. The late phase occurs after some hours and is characterized by an influx of neutrophils, eosinophils and Th2 lymphocytes into the airways (Effros 2007).

Bronchial asthma is a chronic inflammatory airway disease that is defined by reversible airway obstructions and non-specific airway hyper-responsiveness (AHR) (Umetsu 2002).

It is generally accepted that pathological changes in asthma are induced by a chronic inflammatory process which is characterized by infiltration of lymphocytes and eosinophils into bronchial mucosa, increased mucus production and submucosal edema (Hamelmann 2001).

Activated Th2 cells in an atopic individual secrete IL-4, IL-5, IL-9 and IL-13. The increase in IL-2, IL-4 and IL-13 stimulates the production of IgE by B lymphocytes (plasma cells). IgE can bind tightly to FcεR1 receptors on mast cells which release vasoactive amines and cytokines when antigens crosslink these bound IgE molecules. Class switches to IgE cannot occur without IL-4 or IL-13. IL-5 is a so-called hematopoetic cytokine that stimulates eosinophil production in the bone marrow as well as activation and chemotaxis of eosinophils and basophils (Simon 1999). Eosinophils can bind to IgE opsonized surfaces and may secrete major basic and major cationic proteins which can damage lung surfaces (or damage the tough cuticula of helminths) (Effros 2007). IL-9 seems to exert effects on production, activation and chemotaxis regarding mast cells (Renauld 1995, Godfraind 1998).

Within allergic disorders, atopic dermatitis (A.D.) is described as a chronic inflammatory disease of the skin associated with cutaneous hyperreactivity (Leung 2003). The concept of the pathobiology of atopic dermatitis involves a systemic Th2 response, in addition to a biphasic T cell response in the skin, with Th2 cells in acute A.D. and Th1 cells in the chronic phase. Atopic dermatitis involves complex interactions between resident and infiltrating cells influenced by a variety of proinflammatory cytokines and chemokines (Leung 2004).

In contrast, allergic airway hyperreactivity is a chronic disease in which continuously strong Th2 inflammatory responses are induced leading to recruitment of inflammatory cells into the lung and pathological changes of the lung tissues (Umetsu 2002). Type 2 mediated inflammation is characterized by extensive mast cell degranulation and invasion of eosinophils and basophils into the tissue.

Instead of preventing allergies by avoiding risk factors, recent studies try the reverse approach. The idea is to facilitate exposure to protecting factors which may promote induction of immunological tolerance against harmless antigens. In other words the strategy is shifting from allergy prevention by avoidance to tolerance induction (Hamelmann 2008).

Apart from specific immunotherapy (SIT), also application of food-borne microbes and even helminths and their components might be promising candidates.

With all the knowledge available about suppressive capacities of the single recombinant helminth protein Av17 *in vitro*, this study should evaluate if Av17 would be able to modulate two models of allergic disease, namely allergic asthma and atopic dermatitis. Furthermore, gastrointestinal (GI) worm infection should be evaluated in both models, yielding information as to whether or not GI nematode induced immunomodulation is restricted to mucosa-associated disease.

3. Results

3.1 Av17 and Allergic Airway Hyperreactivity (AHR) – a murine asthma model to reveal in vivo effects

3.1.1. Application of filarial cystatin blocks allergic airway disease in a preventive and a pre-challenge approach

In order to study immunomodulatory effects of filarial cystatin (Av17) *in vivo*, Av17 was applied in a murine model for human allergic asthma, namely in the murine model of ovalbumin (OVA) - induced allergic airway hyperreactivity. To induce allergic airway inflammation mice were sensitized twice with the model allergen OVA and subsequently challenged with OVA via the airways.

The capacity of cystatin to alter inflammation was tested in two different settings: the preventive and the pre-challenge approach. For preventive treatment BALB/c mice were injected intraperitoneally with doses of 20µg *E. coli*-expressed recombinant cystatin from *Acanthocheilonema viteae* (Av17) 4 times in weekly intervals. The preventive treatment provides the possibility to interfere with sensitization right from the start: The parasite protein was applied during sensitization together with the first OVA treatment (Fig. 3A). In contrast, pre-challenge treatment with Av17 reveals modulation of inflammation after the sensitization phase with the allergen is completed. Three 20 µg doses of cystatin were applied after the sensitization phase with OVA but before intranasal allergen challenge (Fig. 3B).

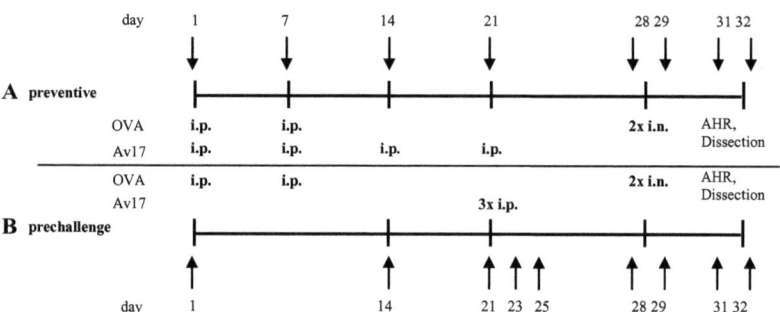

FIGURE 3: Scheme of (**A**) the preventive and (**B**) the pre-challenge model of airway hyperreactivity (AHR). Intraperitoneal applications of ovalbumin (OVA) or filarial cystatin (Av17) are indicated by i.p., intranasal ovalbumin challenge is indicated by i.n.

3.1.2. Filarial cystatin inhibits the development of allergic airway hyperresponsiveness in preventive and pre-challenge setting

Both approaches tested revealed a strong suppression of allergic inflammation in the experimental animals. This was reflected by a downmodulation of hallmarks of murine airway inflammation.

First, significantly reduced total numbers of cells ($p < 0.028$ prev, $p < 0.02$ pre) in bronchoalveolar lavage fluid (BALF) were observed. Total cell numbers assessed were at the level obtained in naïve mice. This effect of cystatin application was most pronounced for eosinophils ($p < 0.05$ prev, $p < 0.028$ pre), which reflects a strong suppression of the main inflammatory cells in the AHR model. Other cell types in BALF did not differ significantly. Treatment with the irrelevant *E. coli*-expressed recombinant control protein DHFR did not lead to significant changes in cell numbers or cell types in the BALF (Fig. 4).

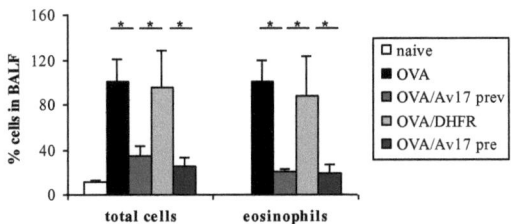

FIGURE 4: Influence of treatment with filarial cystatin (Av17) on total cell numbers and eosinophil numbers in the BALF of mice observed in the preventive (prev) and the pre-challenge (pre) model, respectively. naive: PBS-treated mice; OVA: ovalbumin-treated mice; OVA/Av17: ovalbumin and Av17–treated mice; OVA/DHFR: ovalbumin and DHFR (control protein)–treated mice. Representative data of three individual experiments with 5-6 animals per group; to directly compare data of the pre-challenge with the preventive model data is expressed in percentages (OVA-group is 100%); *$p < 0.05$.

Second, histological analysis of the lung alveolar tissue revealed clear differences between cystatin-treated animals and controls. Only background levels of inflammatory cell infiltration within the lung tissue were observed in hematoxylin/eosin stainings of bronchiole regions of lungs in the preventive model. Furthermore mucus production by goblet cells was found to be nearly absent in mice co-treated with OVA/cystatin but dramatically increased in OVA-controls, as indicated by PAS/H staining for mucin (Fig. 5).

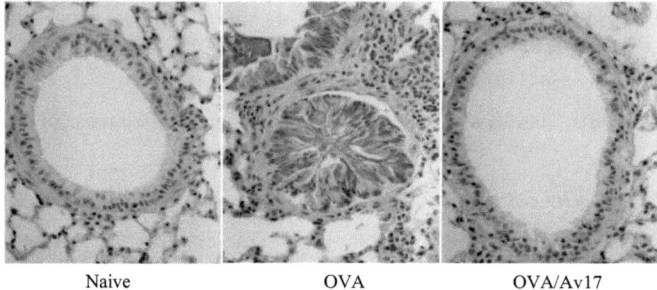

Naive   OVA   OVA/Av17

FIGURE 5: Representative lung sections of Av17-treated mice in the preventive model. Lung sections are stained with PAS/H for analysis of mucus production (red). Increased cell infiltration is indicated in blue (nuclei, he staining). Magnification 400x, histology was performed by Dr. W. Bleiß und A. Marko. naive: PBS-treated mice; OVA: ovalbumin-treated mice; OVA/Av17: ovalbumin and Av17–treatment.

Third, cystatin treatment significantly inhibited the development of *in vivo* airway hyperreactivity (AHR) in treated mice in the preventive ($p < 0.028$) (Fig. 6A) as well as in the pre-challenge model ($p < 0.028$, Fig. 6B). This was indicated by reduced Penh values (Pause enhanced) assessed by whole body plectysmography. Animals co-treated with cystatin and OVA developed less bronchoconstriction in response to increasing metacholin concentrations and therefore had shorter "breaks" (Penh) in breathing.

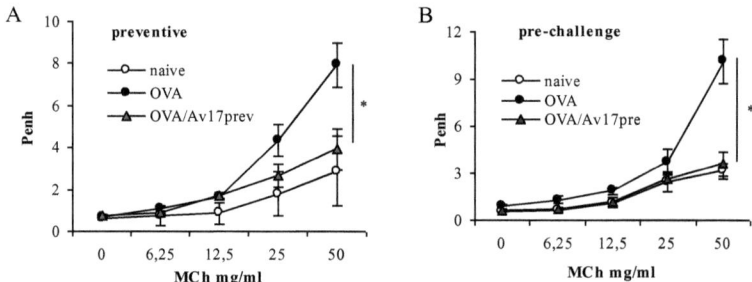

FIGURE 6: Measurement of airway hyperreactivity (AHR) after application of different dosages of methacholine (MCh) to mice treated with filarial cystatin (Av17) (A) in the preventive (prev) and (B) in the pre-challenge (pre) model. AHR is diplayed by Penh (pause enhanced) values. naive: PBS-treated mice; OVA: ovalbumin-treated mice; OVA/Av17: ovalbumin and Av17–treatment. Representative data of two individual experiments with 5-6 animals per group. *$p < 0.05$.

Another typical feature of allergic asthma is the production of allergen-specific antibodies and antibodies of the subclass IgE. Treatment with cystatin significantly reduced serum levels of total IgE (p < 0.0003, Fig. 7A) as well as of OVA-specific IgE (p < 0.0002, Fig. 7B) when applied in the preventive model. This effect was specific to IgE, as serum levels of OVA-specific IgG1 and IgG2a were not significantly altered compared to sensitized and challenged controls (Fig. 7C, D).

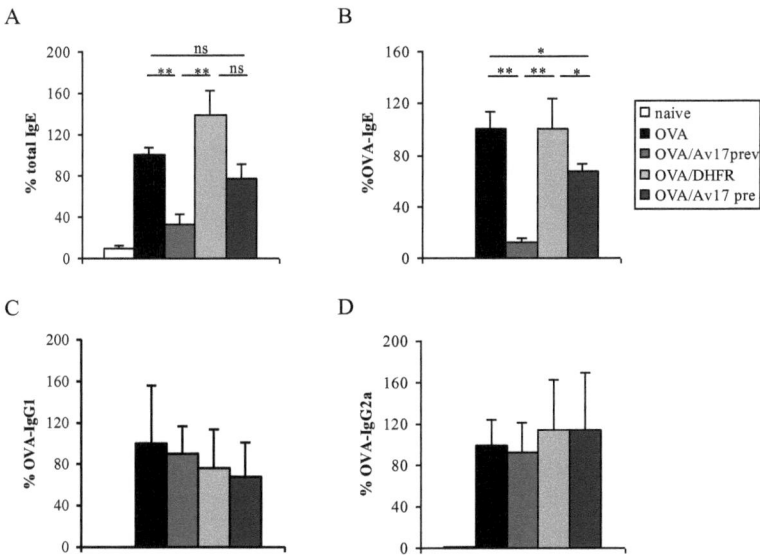

FIGURE 7: Effects of preventive (prev) and pre-challenge (pre) application of filarial cystatin on sensitization were analyzed by measurement of (A) total IgE, (B) OVA-specific IgE, (C) OVA-specific IgG1 and (D) OVA-specific IgG2a concentrations in sera. naive: PBS-treated; OVA: ovalbumin-treated; OVA/Av17: ovalbumin and Av17–treated; OVA/DHFR: ovalbumin and DHFR–treated mice. To directly compare data of the pre-challenge with the preventive model data are expressed in percentages (OVA-group is 100%); Representative data of three individual experiments with 5-6 animals per group. *p < 0.05, **p< 0.005.

Cystatin application in the pre-challenge model resulted in significantly decreased OVA-IgE levels (p < 0.02; Fig. 7B), whereas total IgE production showed only a trend towards down-regulation (Fig. 7A). Hence, in the pre-challenge model cystatin is able to inhibit levels of OVA-specific IgE production following intranasal challenge, whereas amounts of OVA-IgE produced in response to sensitization remain unaffected. OVA – IgG1 and OVA – IgG2a were not altered.

The significant inhibition of IgE production in the preventive model was accompanied by a reduced capacity of sera from cystatin/OVA-treated mice to induce degranulation of basophils. In the RBL-test cells were sensitized with test sera and crosslinking of bound IgE was stimulated with 5µg/ml OVA. OVA/Av17 - treated mice released significantly lower amounts of the mediator ß-hexosaminidase (15 %, p < 0.009) in the preventive approach. This means that less RBL cells degranulated because there was less crosslinking to induce degranulation and less OVA-specific IgE. Interestingly, crosslinking and degranulation effectivity seemed to be downmodulated in pre-challenge approach as well, indicated by reduced ß-hexosaminidase release, although this did not reach statistical significance (Fig. 8). Of interest, no Av17-specific IgE could be measured via RBL test and ELISA (data not shown).

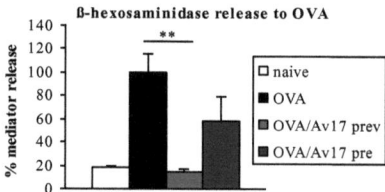

FIGURE 8: Functional OVA-specific IgE was tested in RBL (rat basophil leukemia cell) assay. Degranulation of cells is shown by percentage of mediator release. naive: PBS-treated mice; OVA: ovalbumin-treated mice; OVA/Av17: ovalbumin and Av17–treatment in preventive (prev) and pre-challenge (pre) approach. Representative data of three individual experiments with 5-6 animals per group. **p < 0.005.

Together, these data suggest that cystatin interferes with the recruitment of inflammatory cells to the lung, airway hyperresponsiveness, mucus production in lungs and with allergen-specific IgE production, thus inhibiting the main features of allergen-induced alterations in this mouse model both when applied during and after sensitization.

3.1.3. Filarial cystatin alters cytokine levels

Regarding cytokine patterns the focus was on systemic and local alterations, meaning cytokine levels in splenocyte cultures and BAL-fluid (BALF). Mice treated with OVA/cystatin contained clearly less IL-4 in BALF as animals treated with OVA only (p < 0.02), or with OVA/DHFR (p < 0.028).

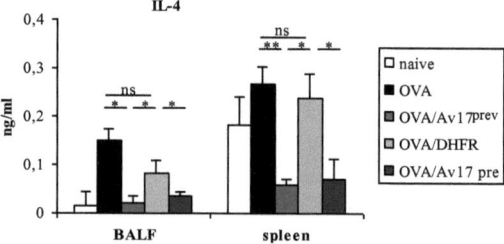

FIGURE 9: Influence of preventive (prev) and pre-challenge (pre) treatment with filarial cystatin on IL-4 levels in bronchoalveolar lavage fluid (BALF) and in splenocytes stimulated with OVA . naive: PBS-treated; OVA: ovalbumin-treated; OVA/Av17: ovalbumin and Av17–treated; OVA/DHFR: ovalbumin and DHFR–treated mice. Representative data of three individual experiments with 5-6 animals per group. *p < 0.05., **p < 0.005

The reduction in this typical Th2 cytokine that is associated with antibody subclass switching to IgE, was significant both in the preventive and in the pre-challenge model (Fig. 9). Along with the local change in BALF, splenocyte cultures of OVA/cystatin-treated mice too, produced significantly less IL-4 in response to allergen stimulation compared to controls in both models ($p < 0.0004$ prev, $p < 0.015$ pre, Fig. 9) accounting for a systemic cystatin effect. Although the levels of IL-10, a strongly anti-inflammatory Th2 cytokine, in BALF were similar between the groups, IL-10 production by spleen cells stimulated with cystatin *ex vivo* was significantly increased in the preventive model ($p < 0.01$) and even more pronounced in the pre-challenge model ($p < 0.002$ compared to OVA, $p < 0.028$ compared to Av17 prev, Fig. 10).

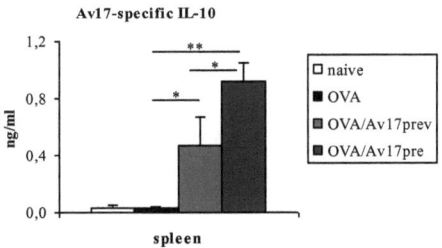

FIGURE 10: Induction of IL-10 production in splenocytes after stimulation with Av17 in the preventive (prev) and pre-challenge (pre) model. naive: PBS-treated; OVA: OVA-treated mice; OVA/Av17: ovalbumin and filarial cystatin–treated mice. Representative data of 3 individual experiments with 4-6 animals are shown; *$p < 0.05$; **$p < 0.005$.

In several experiments cytokine bead arrays were performed to gain further information on cytokines such as eosinophilia-associated IL-5 and mucus-associated IL-13 as well as the Th1 cytokines IL-12, IFN-γ and TNF-α. They were analysed in BALF and in culture supernatant of OVA-stimulated spleen cells (Tab. 1). Both IL-5 and IL-13 levels showed a trend towards reduction at the site of inflammation, in BALF, though the differences did not reach statistical significance. The Th1 cytokines showed similar or slightly decreased amounts in BALF compared to the OVA-group but again none of these changes was significant. Regarding systemic production, both IL-5 and IL-13 levels were elevated similarly to OVA-group, indicating that only parts of the overall Th2-response to OVA were altered systemically by cystatin. However, it could be excluded that the protective effect of Av17 treatment was a mere switch to a Th1-response, because none of the measured Th1 cytokines was elevated in OVA-stimulated splenocytes of cystatin treated animals. Also stimulation with Av17 did not lead to cytokine production other than IL-10.

BALF (pg/ml)	naive	OVA	OVA/Av17
IL-5	48 +/- 9	315 +/- 17	189 +/- 49
IL-13	174 +/- 74	453 +/- 20	273 +/- 48
IFN-γ	22 +/- 7	38 +/- 17	15 +/- 0,1
TNFα	63 +/- 9	76 +/- 13	54 +/- 0,1
IL-12p40	30 +/- 3	90 +/- 44	48 +/- 8

spleen (pg/ml)	naive	OVA	OVA/Av17
IL-5	529 +/- 132	2746 +/- 45	2833 +/- 162
IL-13	1280 +/- 612	2676 +/- 806	2578 +/- 469
IFN-γ	2826 +/- 450	85 +/- 32	154 +/- 133
TNFα	127 +/- 36	278 +/- 6	284 +/- 10
IL12p40	315 +/- 11	87 +/- 28	150 +/- 8

Table 1: Cytokine levels determined in bronchoalveolar lavage fluid (BALF) and in OVA-stimulated splenocytes by cytokine bead array. Th2 cytokines IL-5 and IL-13 as well as Th1 cytokines IFN-γ, TNFα and IL-12p40 from an individual experiment with 5 animals per group in the pre-challenge setting is shown. Naïve: PBS-treated animals, OVA: ovalbumin-treated animals, OVA/Av17: ovalbumin and filarial cystatin-treated animals.

TGF-ß values were assessed in BALF and determined by real-time PCR in mRNA of lung tissue but again showed a trend towards reduction but no significant alteration in mice treated with cystatin compared to controls (data not shown). Finally proliferation of spleen cells in response to OVA was monitored to rule out mere anergy of splenocytes resulting in lower cytokine levels. However, proliferation was never significantly altered in cystatin-treated animals compared to control groups (data not shown).

These data are compatible with a strong down-regulation of effector functions of the allergic immune reaction by filarial cystatin, resulting in abrogation of allergic sensitization and impaired Th2 immune responses, reduced eosinophilic airway inflammation and decreased airway hyperreactivity.

3.1.4. Filarial cystatin targets macrophages

In previous studies, it was shown that filarial cystatin targets and alters macrophages *in vitro* (Hartmann 1997) leading to IL-10 production and T cell suppression. Therefore, the relevance of macrophages in the mouse model of OVA-induced airway inflammation was investigated. Macrophages were selectively depleted *in vivo* by use of clodronate containing liposomes (MLV) (van Roijen 1994). These multilamellar vesicles are made of lipids selecting for uptake by macrophages. In addition the chemical is not phagocytosed when "free" outside the vesicles, preventing unspecific killing of phagocytes by release of clodronate of killed macrophages. Clodronate MLVs were applied intraperitoneally and intranasally two days prior to airway allergen challenges of sensitized and treated mice. This treatment led to the loss of > 95% of macrophages in the BALF and in the peritoneum, as confirmed by FACS staining of cells positive for the macrophage marker F4/80 and negative for the B cell marker CD19 and the T cell marker CD3 (Fig. 11). Additional analyses confirmed that macrophages remained absent for at least 6 days (data not shown).

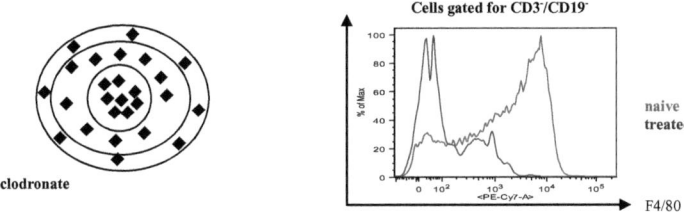

FIGURE 11: Depletion of macrophages is achieved by application of multilamellar vesicles (MLV) filled with the chemical clodronate. Schematic view of MLV and representative FACS plot of peritoneal excudate cells (PEC) 4 days after application: Cells are gated for no expression of T cell marker CD3 (CD3-) and B cell marker CD19 (CD19-). Macrophages are identified by expression of the marker F4/80. naïve (red): no MLV-application, treated (blue): application of MLV in Av17-treated animals.

To rule out effects of macrophage depletion on the model OVA-treated mice were depleted of macrophages in a preliminary test. Depletion of macrophages neither prevented nor enhanced the onset of the allergic airway inflammation, indicated by the absence of significant changes on the cell recruitment into the lung within the OVA-group (Fig. 12). In contrast, in cystatin-treated animals the depletion of macrophages led to a reversion of the total cell numbers (p < 0.008) and eosinophils (p < 0.002) in BALF to the level of the OVA-sensitized and challenged control group (Fig. 12).

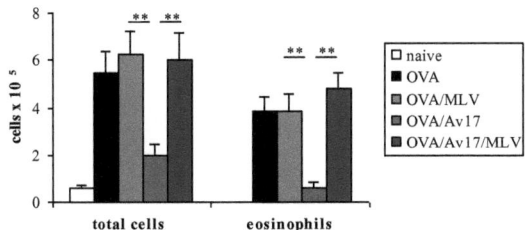

FIGURE 12: Effects of filarial cystatin are dependent on macrophages. Shown are the total cell numbers and eosinophil numbers in BALF in mice depleted of macrophages in the preventive cystatin-treatment approach. naive: PBS-treated mice; OVA: ovalbumin-treated mice; OVA/MLV: ovalbumin and MLV-treated mice; OVA/Av17: ovalbumin and filarial cystatin (Av17)-treated mice; OVA/Av17/MLV: ovalbumin and Av17-treated mice depleted of macrophages by MLV; Representative data of three individual experiments with 5-6 animals per group **p < 0.005

Interestingly, the treatment with multilamellar vesicles filled with clodronate also significantly elevated the production of total IgE (p < 0.05, Fig. 13A) and of OVA-specific

IgE in sera of OVA/cystatin–treated mice (p < 0.007; Fig. 13B). However, antibody levels were still lowered in comparison to OVA-control group.

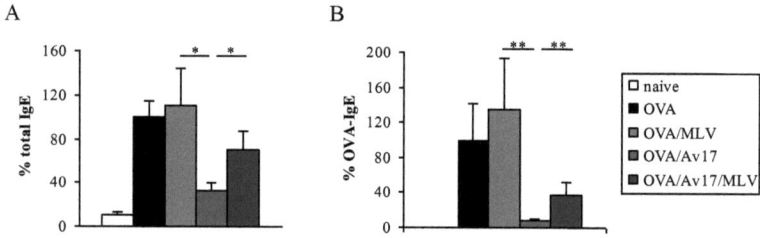

FIGURE 13: Influence of macrophage depletion on (A) total IgE and (B) OVA-specific IgE levels in sera of treated mice (preventive approach). naive: PBS-treated mice; OVA: ovalbumin-treated mice; OVA/MLV: ovalbumin and MLV-treated mice; OVA/Av17: ovalbumin and filarial cystatin (Av17)-treated mice; OVA/Av17/MLV: ovalbumin and Av17-treated mice depleted of macrophages by MLV; IgE values are expressed in percentages (OVA = 100%). Representative data of three individual experiments with 5-6 animals per group. *p < 0.05;**p < 0.005.

OVA-specific IL-4 production of restimulated spleen cells was not significantly altered by macrophage depletion, although a trend towards upregulation could be observed (Fig. 14A). However, AHR of macrophage-depleted, OVA/cystatin–treated mice rose significantly (p < 0.04) (Fig. 14B).

Interestingly, macrophage depletion had a drastic effect on IL-10 production of splenocytes. Lack of macrophages resulted in a significant decrease of IL-10 production in Av17-stimulated cultures indicating an important role for macrophages in Av17-specific IL-10 production (Fig. 14C).

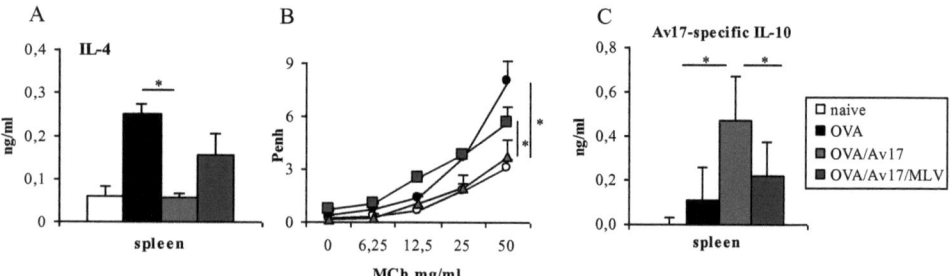

FIGURE 14: Effects of macrophage depletion on (A) OVA-specific IL-4 production of spleen cells and (B) airway hyperreactivity of mice after treatment with different dosages methacholine (MCh). (C) Av17-specific IL-10 production in splenocytes. Naive: PBS-treated mice; OVA: ovalbumin-treated mice; OVA/Av17: ovalbumin and filarial cystatin (Av17)-treated mice; OVA/Av17/MLV: ovalbumin and Av17-treated mice depleted of macrophages; Penh: pause enhanced, Representative data of three individual experiments with 5-6 animals per group.*p < 0.05.

Together, these results indicate that macrophages translate the immunomodulatory effects of cystatin airway inflammation, airway hyperreactivity and IgE production and are strongly involved in IL-10 production.

3.1.5. Cystatin increases the number of regulatory T cells

As T regulatory cells have been described to play an important role in allergy and asthma (Hawrylowicz 2005a) we aimed to determine whether cystatin treatment had an influence on the numbers of Treg cells at the site of inflammation. Hence, we analyzed the co-expression of the surface markers CD4, CD25 and CD103 on peribronchial lymph node cells (PBLNC). The proportion of Treg cells was significantly elevated in OVA/cystatin-treated animals as compared to OVA-controls (3% versus 1.9%, $p < 0.05$) and to OVA/DHFR-controls (3% versus 2.1%; $p < 0.05$; Fig. 15).

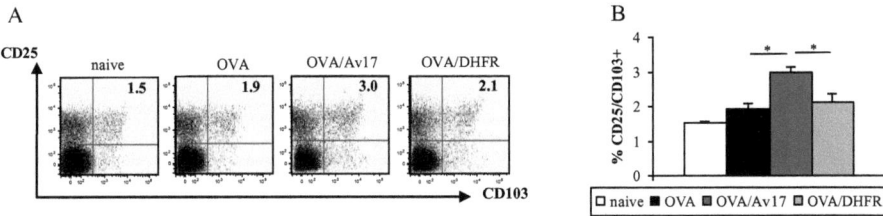

FIGURE 15: Treatment of mice with filarial cystatin in the preventive approach alters Treg numbers. Shown are (A) FACS plots of PBLN cells of one representative animal per group and (B) numbers of Treg cells ($CD4^+CD25^+CD103^+$) in PBLN as mean of groups; naïve: PBS-treated mice; OVA: ovalbumin-treated mice; OVA/Av17: ovalbumin and (Av17)-treated mice, OVA/DHFR: ovalbumin and DHFR–treated mice; Representative data of three individual experiments with 5-6 animals per group *$p<0.05$

To corroborate the identification of Treg cells, in several experiments the transcription factor forkhead box transcription factor p3 (Foxp3), which is regarded as one of the most reliable Treg markers at the moment, was stained in addition to the cell surface markers. More than 96% of the Treg cells expressed Foxp3 (Fig. 16A) verifying the analyses.

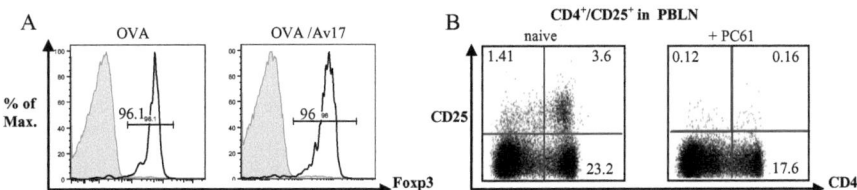

FIGURE 16: (A) Representative histograms of Foxp3$^+$ cells in $CD4^+CD25^+CD103^+$ positive PBLNC. Histograms are representative for all animals in all groups. OVA: ovalbumin-treated mice; OVA/Av17: ovalbumin and Av17-treated mice. (B) Verification of Treg - depletion effectivity. Shown are representative plots of CD4/CD25 gated PBLN (peribronchial lymph node cells) without (left) and after (right) αCD25Ab (PC61) treatment. Naïve: no PC61 treatment, + PC61: after application of αCD25Ab

In order to analyze the role of Treg cells in cystatin-induced immunomodulation sensitized animals were treated with anti-CD25 antibodies five days prior to the first airway allergen challenge. Application of PC61 (αCD25Ab) significantly decreased the number of $CD4^+CD25^+CD103^+$ Treg cells to 0.16% in PBLNC (Fig. 16B).

Analysis of cell numbers and types in BALF of cystatin treated mice with depleted Treg cells was performed: The levels of total cells as well as eosinophils showed a trend of restoration though this did not reach statistical significance (Fig. 17). Application of isotype-matched control antibodies did not alter cystatin effects on BALF cells.

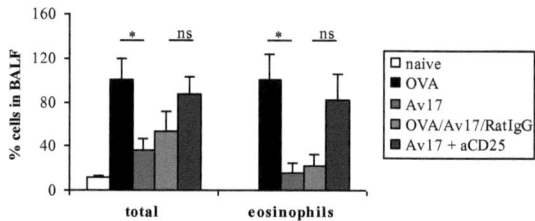

FIGURE 17: Depletion of CD25-positive cells partly reversed the immunomodulation exerted by prevetive treatment with filarial cystatin. Shown are alterations in total cell and eosinophil numbers in BALF. naive: PBS-treated mice; OVA: ovalbumin-treated mice; OVA/Av17: ovalbumin and filarial cystatin (Av17)-treated mice; OVA/Av17/rat IgG: mice treated with OVA and Av17 plus isotype-matched control antibodies, OVA/Av17/αCD25: OVA and Av17-treated mice depleted of Treg cells by application of αCD25 antibodies, Representative data of 3 individual experiments with 4 animals per group. *p < 0.05.

However, the production of total IgE (p < 0.002; Fig. 18A) and OVA-specific IgE (p < 0.002; Fig. 18B) was significantly restored compared to cystatin treated animals with non-manipulated Treg cells.

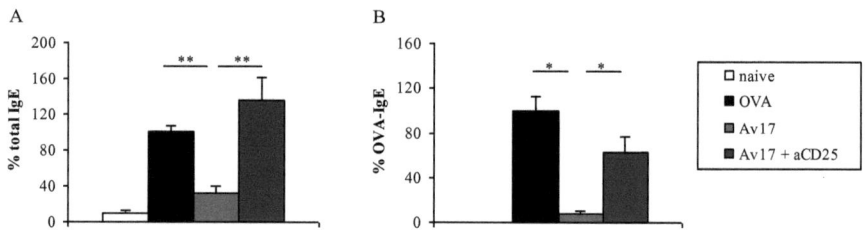

FIGURE 18: Depletion of Tregs in filarial cystatin-treated (Av17) mice in the preventive model affects levels of (A) total and (B) OVA-specific serum IgE; naive: PBS-treated mice; OVA: ovalbumin-treated mice; OVA/Av17: OVA and (Av17)-treated mice; OVA/Av17/αCD25: OVA and Av17-treated mice depleted of Tregs, IgE-values are expressed in percentages (OVA-group is 100%); Representative data of 3 individual experiments with 4 animals per group *p < 0.05, **p < 0.005.

Anyhow, no changes were determined for allergen-specific IL-4 production (Fig. 19A) and development of AHR in cystatin-treated mice after depletion of Treg cells (Fig. 19B).

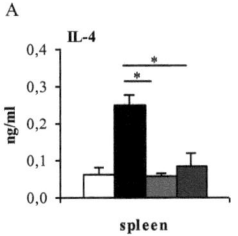 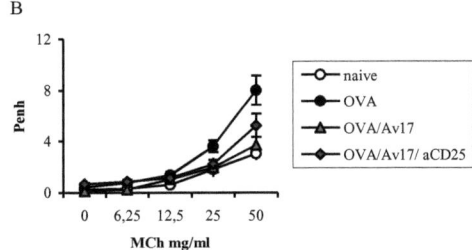

FIGURE 19: Effect of depleting Tregs in Av17 preventive model on (A) OVA-specific IL-4 production in spleen and on (B) airway hyperreactivity in response to increasing metacholine (MCh) concentrations. naive: PBS-treated mice; OVA: ovalbumin-treated mice; OVA/Av17: OVA and filarial cystatin (Av17)-treated mice; OVA/Av17/aCD25: OVA and Av17-treated mice depleted of Tregs. Representative data of 3 individual experiments with 4 animals per group. *$p < 0.05$.

These data indicate that Treg cells are involved in the cystatin-induced effects on murine airway hyperreactivity, albeit to a less prominent degree than macrophages.

3.1.6. Inhibition of allergic responses by cystatin is dependent on IL-10

When addressing the question of possible mediators for the protective effect of cystatin in the murine asthma model IL-10 seems to be a striking candidate for several reasons. First, as mentioned above, previous *in vitro* studies had shown that cystatin is able to induce the production of anti-inflammatory IL-10 by macrophages. Second, the depletion of macrophages *in vivo* reversed the effects of cystatin in the asthma model. And third, IL-10 production was profoundly inhibited in animals lacking macrophages and not being protected from asthma (Fig. 14C). Therefore it was hypothesized that IL-10 might be a key mediator of cystatin-induced immunomodulation. To analyze its influence, antagonizing anti-IL-10 receptor antibodies (anti-IL-10R Ab) were used in the pre-challenge model of OVA-induced airway hyperreactivity. The pre-challenge model was chosen because Av17-specific IL-10 production in spleen cells was even more pronounced in this approach in comparison to the preventive approach (Fig. 10).

Anti-IL-10R was injected 3 times along with the filarial cystatin after sensitization and prior to the first allergen airway challenge with OVA. Blocking of the IL-10R in OVA/cystatin treated animals completely abrogated the protective effect of Av17. Total cell numbers in BALF were restored to the degree observed in sensitized and challenged positive control animals ($p < 0.02$) (Fig. 20). This effect of blocking IL-10R was most pronounced for the eosinophilic airway infiltration ($p < 0.02$) (Fig. 20).

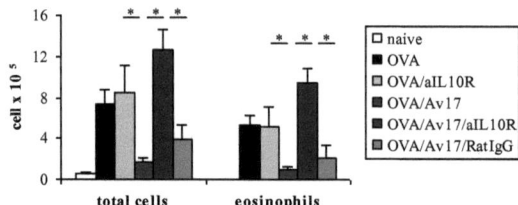

FIGURE 20: Suppression of allergic responses by filarial cystatin is dependent on IL-10. Shown is the effect of application of aIL10R Ab to mice in the pre-challenge model on total cell numbers and eosinophil numbers; naive: PBS-treated mice; OVA: ovalbumin-treated mice; OVA/aIL-10R: mice treated with OVA and anti-IL-10 receptor antibodies; OVA/Av17: OVA and filarial cystatin (Av17)–treated mice; OVA/Av17/aIL-10R: mice treated with OVA, Av17 and anti-IL-10R-Ab; OVA/Av17/rat IgG: mice treated with OVA, Av17 and isotype-matched control antibodies. Representative data of 2 individual experiments with 6 animals per group. *$p < 0.05$.

Similarly, the development of AHR ($p < 0.048$) (Fig. 21A), and the production of OVA-specific IgE were increased to the levels observed in OVA-sensitized and challenged control mice ($p < 0.03$) (Fig. 21B). However, the inhibition of allergen-specific IL-4 production in OVA/cystatin treated animals was not altered after application of anti-IL-10R antibodies (Fig. 21C). Application of isotype matched control antibodies had no effects on the cystatin-induced parameters as well as application of αIL-10R did not alter the values of the OVA positive control group.

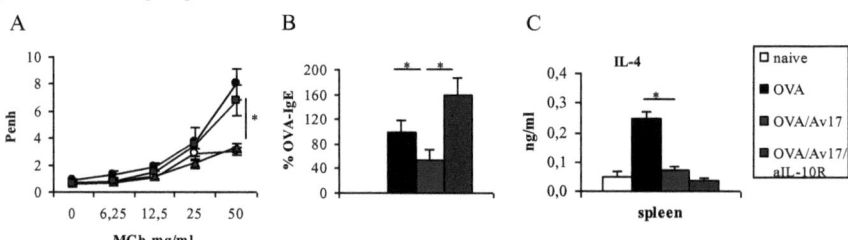

FIGURE 21: Effects of αIL10R-Ab application in the pre-challenge filarial cystatin approach on (A) airway hyperreactivity, (B) levels of OVA-specific serum IgE and (C) OVA-specific IL-4 production of spleen cells. naive: PBS-treated mice; OVA: ovalbumin-treated mice; OVA/Av17: OVA and filarial cystatin (Av17)–treated mice; OVA/Av17/aIL-10R: mice treated with ovalbumin, Av17 and anti-IL-10 receptor antibodies; OVA-IgE values are expressed in percentages (OVA-group is 100%); Representative data of 2 individual experiments with 6 animals per group. *$p < 0.05$

These data indicate that IL-10 is a key cytokine in filarial cystatin-induced modulation of allergic disease, although IL-10 independent mechanisms like the suppression of allergen-specific IL-4 have to be taken into account too.

3.1.7. Macrophages are main producers of Av17-specific IL-10

Macrophages and Treg cells are both potent sources of IL-10. To unravel which of these cells would be primarily responsible for the IL-10 production after treatment with filarial cystatin splenocytes of differently treated animals were stimulated with cystatin. Cystatin-treated animals showed significantly increased levels of IL-10 in spleen in comparison to OVA-treated animals ($p < 0.002$ pre-challenge model; $p < 0.01$ preventive model, $p < 0.024$ comparison of pre-challenge to preventive model; Fig. 10). However, after depletion of macrophages, the IL-10 production was significantly decreased in the OVA/cystatin-treated animals ($p < 0.04$; Fig. 22), whereas such an effect was not observed in spleens of animals depleted of Treg cells. Treg–depleted mice actually showed a trend of elevated IL-10 values (Fig. 22).

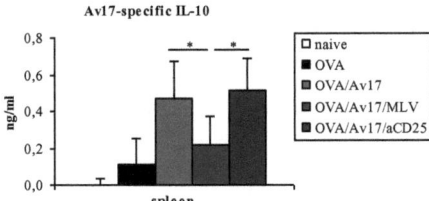

FIGURE 22: Monitoring IL-10 production in response to Av17 stimulation in spleen cells after depletion of macrophages or Tregs in the preventive model; naive: PBS-treated mice; OVA: ovalbumin-treated mice; OVA/Av17: OVA and Av17–treated; OVA/Av17/MLV: OVA and Av17-treated mice depleted of macrophages; OVA/Av17/aCD25: OVA and Av17-treated mice in which Treg cells were depleted. Representative data of 3 individual experiments with 4-6 animals per group. *$p < 0.05$.

These data underline the pivotal role of macrophages in the cystatin-induced modulation of allergic airway inflammation and hyperreactivity.

3.1.8. Features of Av17 application in vivo outside inflammation settings

To rule out negative effects of cystatin on healthy individuals and find out more about cystatin-dependent effects on the immune system, cystatin was applied 4 times to BALB/c mice in weekly intervals (20µg doses) without allergen-sensitization and challenged with either the immunomodulator Av17 (Av17ip/Av17in) or the model allergen OVA (Av17ip/OVAin) via the airways. Assessment of airway reactivity and Penh values clearly ruled out any negative effect of cystatin on these parameters. Penh values of Av17ip/in and Av17ip/OVAin groups were on naïve levels or even showing a trend of reduction, regarding Av17ip/in group (Fig. 23A).

Results

Analyzing cell numbers and cell types in BALF corroborated these data, although intranasal challenge with proteins instead of PBS (as in the naïve group) led to influx of small numbers of cells. However the influx was far from scenarios found in airway inflammation (Fig. 23B).

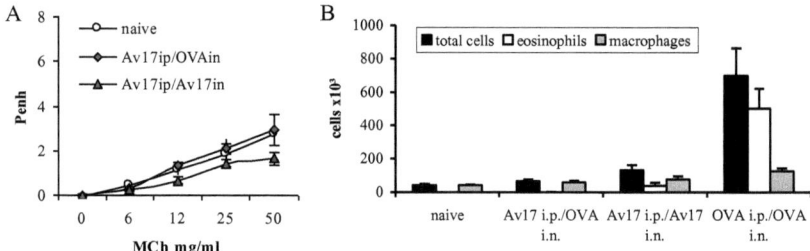

FIGURE 23: (A) Airway hyperreactivity and (B) cell numbers in animals injected with cystatin but not sensitized with ovalbumin. naïve: PBS-treated mice; Av17ip/OVAin: intraperitoneal (ip) injections of Av17 and intranasal (in) challenge with ovalbumin (OVA); Av17ip/Av17in: intraperitoneal and intranasal Av17 application; OVAip/OVAin: comparison values of OVA ip and OVA in treated mice (no Av17). Shown are mean values of 5 animals per group.

Regarding antibody production cystatin application led to the induction of specific antibodies. However no Av17-specific IgE was measurable. Mostly the subclasses IgG1 (described as competitor to IgE) and IgG/M were detected (Fig. 24).

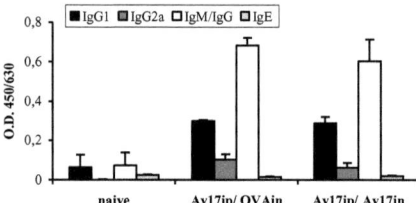

FIGURE 24: Cystatin-specific antibodies in sera of cystatin injected mice. Subclass test for IgG1, IgG2a, IgM/IgG and IgE. Serum levels of cystatin-specific antibodies shown as optical densities (O.D.). Naïve: PBS-treated mice; Av17ip/OVAin: intraperitoneal (ip) injections of Av17 and intranasal (in) challenge with ovalbumin (OVA); Av17ip/Av17in: intraperitoneal and intranasal Av17 application; Shown are mean values of 5 animals per group.

Regarding the IL-10 inducing effect of cystatin on splenocytes and macrophages found in earlier experiments, it was of special interest to survey also peritoneal excudate cells (PECs). First, Av17-stimulated IL-10 secretion of PECs derived from naïve and cystatin-treated mice were compared. Both naïve and Av17-PEC were found to produce high levels of IL-10 when stimulated with cystatin. The Av17-treated PECs produced a little more IL-10 than naïve PECs but this was not statistical significant (Fig. 25A). Cystatin-stimulated splenocytes produced very low levels of IL-10 in this setting (up to 100pg in comparison to up to 400pg in

PECs). In addition, IL-10 production was not found in naïve splenocytes, whereas intranasal treatment seemed to enhance production, although this was not statistical significant (Fig. 25B). These data point to macrophages in PECs as main IL-10 producers.

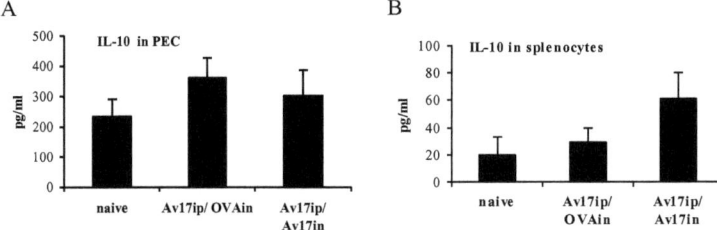

FIGURE 25: Stimulation of (A) peritoneal excudate cells (PEC) and (B) splenocytes derived from naïve or cystatin-treated animals with Av17. Shown are mean values of IL-10 levels after 24h stimulation of 5 animals per group. naïve: PBS-treated mice; Av17ip/OVAin: i.p. injections of Av17 and i.n. challenge with ovalbumin (OVA); Av17ip/Av17in: intraperitoneal and intranasal Av17 application;

To examine IL-10 release in Av17-stimulated Av17-PECs in more detail, the cells were adhered overnight on cell culture plate surfaces and thus enriched for macrophages (by discarding supernatants). When these enriched PEC macrophages were stimulated with Av17 or a mutated form of cystatin (mAv17, no proteinase inhibitor function) a clearly increased secretion of IL-10 was found. In comparison, incubation with medium or the control proteins DHFR (dihydrofolat-reductase) and Cysele2 (*C. elegans* cystatin 2, a cystatin homolog of a non-parasitic helminth) did not induce IL-10 secretion. Furthermore IL-10- secretion in response to Av17 stimulation could not be blocked by polymyxin B (10µg/ml). Interestingly, lipopolysaccharide (LPS) stimulation also induced IL-10-secretion but this effect could be clearly inhibited by addition of polymyxin B (Fig. 26).

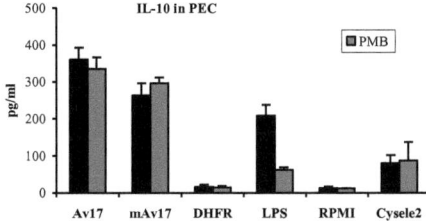

FIGURE 26: Stimulation of enriched PEC derived from cystatin treated animals. Shown is the IL-10 release in response to 24h stimulation with filarial cystatin (Av17), mutated Av17 (mAv17), unrelated control protein (DHFR), lipopolysaccharid (LPS), medium (RPMI) and *C.elegans* cystatin 2 (Cysele2).

These data indicate, that only parasite-derived cystatin leads to IL-10 production, as verified by a lack of IL-10 production with soil-nematode derived cystatin (Cysele2). Furthermore the data suggest that IL-10 production is not due to LPS contamination, as addition of the LPS-ligand polymyxin B could only block LPS-induced IL-10 production but not Av17-induced secretion.

3.2. Av17 in allergic skin-disease: ovalbumin-induced atopic dermatitis

Considering the protective effect of cystatin application in the model of allergic airway inflammation led to the question whether effects of cystatin application would be restricted to the Th2 dependent asthma model or if Av17 would also yield amelioration of disease on another allergy model, namely atopic dermatitis. The dermatitis model is described to consist of a mixed Th2 and Th1 reaction and inflames a different tissue (cutaneous vs mucosal). This approach aimed to provide insight into the "effector range" of cystatin. Furthermore, dermatitis is often claimed to be the start of the atopic march, meaning that a high percentage of dermatitis patients develop asthma, food allergy or other allergic diseases in their later life (Boguniewicz 2004). Inhibiting dermatitis could therefore raise a possibility to stop the atopic march.

3.2.1. Treatment with filarial cystatin prevents eczema in murine atopic dermatitis

To induce atopic dermatitis BALB/c mice were injected intraperitoneally (i.p.) with the model allergen ovalbumin (OVA) and subsequently challenged epicutaneously (e.c.) via OVA-saturated patches. Treatment with purified recombinant cystatin from *A. viteae* was performed during sensitization with OVA and along with the epicutaneous challenge with the allergen (Fig. 25).

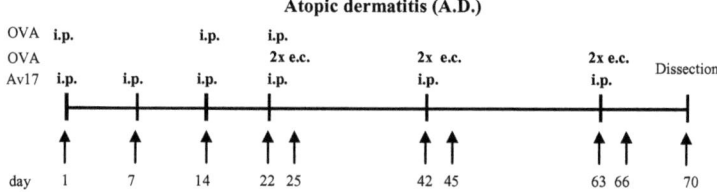

FIGURE 27: Model of murine atopic dermatitis. OVA i.p. indicates sensitization with ovalbumin; OVA e.c. symbols epicutanous treatment with ovalbumin, 2 cycles of 3 days; Av17 i.p. signals intraperitoneal injection of filarial cystatin (Av17).

Treatment with 6 doses of cystatin (20 µg each) but not with the irrelevant recombinant control protein murine dihydrofolate reductase (DHFR) lead to significantly reduced occurrence and severity of eczema. This was evaluated by assessment of a clinical skin score depending on comparison of erythema, edema/papulation, excoriation/crusting, dryness and extension of skin and eczema, respectively. Clinical skin score in sensitized animals that were treated along with cystatin (OVA/Av17-group) showed a mean skin score of 4.2, whereas both control groups had elevated skin scores of 7.4 (OVA, p< 0.04) or 10.8 (DHFR, p<0.008) respectively compared to a skin score of 3.0 in naïve mice (Fig. 28).

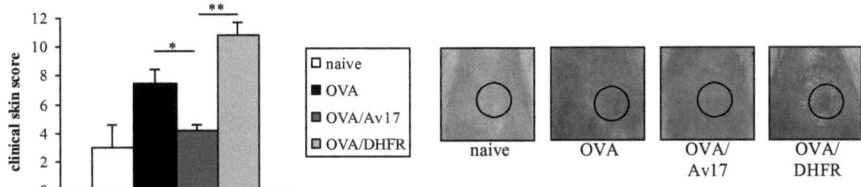

FIGURE 28: Assessment of clinical skin score values in the dermatitis model. Shown are mean values for groups and representative photos of patch areas on the belly (indicated by black circles). naive: i.p. and e.c. PBS-treated mice; OVA: i.p. and e.c. ovalbumin-treated mice; OVA/Av17: i.p. and e.c. OVA-treated mice injected with filarial cystatin; OVA/DHFR: i.p. and e.c. OVA-treated mice injected with control protein DHFR. Representative data of 2 individual experiments with 3-6 animals per group. *$p < 0.05$, **$p < 0.005$.

These findings were underlined by the measurement of thickness of epidermis in eczema/patch regions. Treatment with cystatin led to a decrease in epidermis thickness back to naïve levels (32 µm and 30 µm) in comparison to values of 52 µm and 60 µm in OVA and OVA/DHFR group ($p < 0.001$), respectively (Fig. 29).

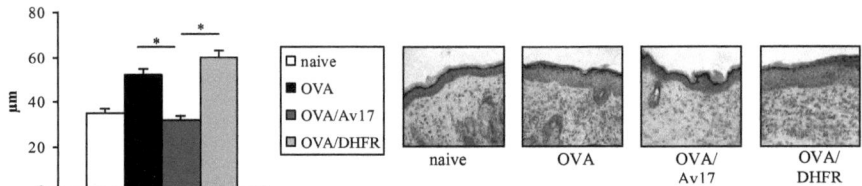

FIGURE 29: Thickness of epidermis was measured in skin sections of patch areas. Shown are mean values for groups and representative pictures of sections stained with he. naive: i.p. and e.c. PBS-treated mice; OVA: i.p. and e.c. ovalbumin-treated mice; OVA/Av17: i.p. and e.c. OVA-treated mice injected with filarial cystatin; OVA/DHFR: i.p. and e.c. OVA-treated mice injected with control protein DHFR. Measurement was performed by A. Dahten. Representative data of 2 individual experiments with 3-6 animals per group. *$p < 0.05$.

Both the clinical skin score and measurement of epidermis thickness clearly revealed that cystatin application does lead to prevention or amelioration of eczema formation, an effect not seen for the control protein.

3.2.2. Cell infiltration into challenged skin is altered by cystatin treatment

In order to obtain information on cell influx into treated tissue, sections of patched skin regions were histologically analysed. Staining for CD4 and CD8 T cells as well as for mast cell and CD11c positive dendritic cell infiltration was performed. Treatment with the immunomodulator led to significantly less infiltration of CD4 helper T cells ($p < 0.02$) and a clear trend of less mast cells compared to controls (Fig. 30A, B).

Results

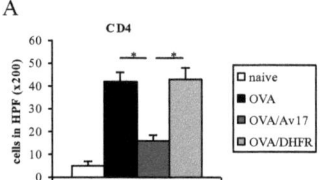
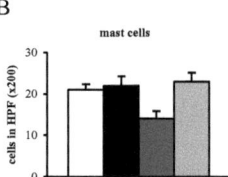

FIGURE 30: Histological skin sections reflecting cell influx. (A) CD4 positive T cells and (B) mast cells were stained in individual sections. Shown are mean values for groups. naive: i.p. and e.c. PBS-treated mice; OVA: i.p. and e.c. ovalbumin-treated mice; OVA/Av17: i.p. and e.c. OVA-treated mice injected with filarial cystatin; OVA/DHFR: i.p. and e.c. OVA-treated mice injected with control protein DHFR. Staining was performed by A. Dahten. Representative data of 2 individual experiments with 3-6 animals per group. *$p < 0.05$.

Interestingly, in these experiments influx of both CD8 positive T cells and CD11c positive dendritic cells was not linked to severity of eczema (data not shown) as both cell types were found also in low levels in DHFR-treated animals, but skin score was dramatically increased in DHFR-groups.

3.2.3. Changes in OVA-specific IgE in sera upon cystatin-treatment

To analyze effects of cystatin treatment on sensitization, allergen specific antibodies were measured according to the analysis in the asthma model. Sera of sensitized and challenged animals treated along with cystatin contained significantly reduced allergen-specific IgE levels ($p < 0.01$) however also the control protein treated group revealed a trend to reduction though this did not reach statistical significance (Fig. 31A).

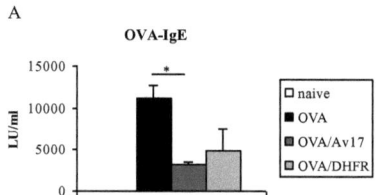
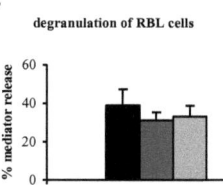

FIGURE 31: Allergen-specific IgE-levels in filarial cystatin-treated animals. Shown are (A) mean values of ovalbumin-specific IgE in LU (lab units) and (B) degranulation of rat basophil cells (RBL) in response to OVA-antigen and sera, indicated by percentage of mediator release. naive: i.p. and e.c. PBS-treated mice; OVA: i.p. and e.c. ovalbumin-treated mice; OVA/Av17: i.p. and e.c. OVA-treated mice injected with filarial cystatin; OVA/DHFR: i.p. and e.c. OVA-treated mice injected with control protein DHFR. Representative data of 2 individual experiments with 3-6 animals per group. *$p < 0.05$.

OVA-IgG1- and OVA-IgG2a- amounts were found to be unaltered by cystatin treatment, though they showed a trend towards down-modulation in comparison to the OVA-group (not shown). Total IgE-concentrations also tended to be decreased but without reaching statistical significance (not shown). In addition, a mediator release test to investigate IgE-activity was performed and resulted in equal activity-levels for all OVA-sensitized and challenged groups not dependent on protein treatment (Fig. 31B). This indicates that there is a downmodulatory effect of cystatin on allergen-specific IgE in the dermatitis model but that reduction of OVA-IgE levels is not the crucial factor for amelioration in this approach.

Therefore, it can be concluded that cystatin treatment decreases the clinical picture of atopic dermatitis and leads to alterations in cell recruitment but this is obviously not dependent on antibody levels in this late phase of the model. This fits to reports that the late chronic phase of dermatitis is Th1-dependent, in contrast to the early acute phase described to be Th2-dependent (Leung 2004).

3.2.4. Cytokine pattern in spleen indicates altered response to the allergen

Systemic cytokine production was analyzed in splenocyte cultures by stimulation of spleen cells with the allergen ovalbumin (OVA). Splenocytes derived from cystatin-treated animals secreted clearly reduced amounts of OVA-specific IL-4, IL-5 and IL-10 although this did not reach statistical significance. However, all three cytokines were found to be up-regulated in control groups (Fig. 32A, B, C). IFN-γ and TNF-α were found to be not induced or the latter even undetectable (data not shown). Stimulation of splenocytes with filarial cystatin revealed upregulated IL-10 secretion in splenocytes derived from cystatin-treated mice as observed in the asthma model (data not shown). This indicates that cystatin-treatment did clearly alter systemic response to the allergen.

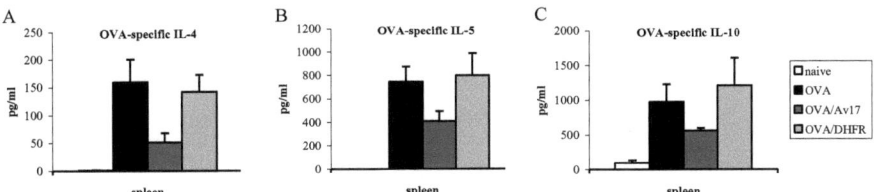

FIGURE 32: Splenocytes of mice in the dermatitis model were stimulated with ovalbumin (OVA) and release of OVA-specific (A) IL-4, (B) IL-5 and (C) IL-10 was measured. naive: i.p. and e.c. PBS-treated mice; OVA: i.p. and e.c. ovalbumin-treated mice; OVA/Av17: i.p. and e.c. OVA-treated mice injected with filarial cystatin; OVA/DHFR: i.p. and e.c. OVA-treated mice injected with control protein DHFR. Representative data of 2 individual experiments with 3-6 animals per group.

3.2.5. Local cytokine production in skin is altered by cystatin treatment

To monitor cytokines directly at the site of inflammation skin samples of the patch region were analyzed via real time PCR. Levels of IL-4-, IL-10- and IFN-γ- mRNA in OVA/Av17 patch region skin compared to OVA- group were decreased (Tab. 2). IL-4 and IL-10 transcripts were elevated in OVA/DHFR group whereas IFN-γ-mRNA was still decreased.

However none of these changes reched statistical significance. Interestingly, an elevation of anti-inflammatory and wound-healing-associated TGF-β-mRNA levels in cystatin co-treated tissue samples compared to OVA- group was detected (p < 0.031) and this increase was about 50% lower in the control-protein co-treated group. Foxp3 levels in skin of Av17-treated animals were not altered (data not shown).

Compared to OVA	Fold change (mean)		
	OVA	OVA/Av17	OVA/DHFR
IL-4	1.04 +/- 0.17	0.33 +/- 0.13	4.22 +/- 1.85
IL-10	1.11 +/- 0.26	0.91 +/- 0.21	4.09 +/- 1.70
TGF-β	2.06 +/- 1.60	17.8 +/- 7.40	7.89 +/- 4.60
IFN-γ	1 +/- 0	0.59 +/- 0.29	0.22 +/- 0.06

Table 2: Skin patch regions were analysed in real time PCR for mRNA of cytokines. Shown are fold changes in local cytokines compared to OVA-group indicated by mean values +/- SEM. Green: down-regulation of mRNA; Red: up-regulation of mRNA; OVA: i.p. and e.c. ovalbumin-treated mice; OVA/Av17: i.p. and e.c. OVA-treated mice injected with filarial cystatin; OVA/DHFR: i.p. and e.c. OVA-treated mice injected with control protein DHFR.

3.2.6. Cystatin restores Treg numbers in mesenteric lymph nodes but does not alter inguinal lymph nodes

As there are implications that Treg cells may play a role in atopic dermatitis the skin-draining lymph node cells (inguinal lymph node cells, ILN) and unrelated mesenteric lymph node cells (MLN) were analyzed for the expression of the surface markers CD4, CD25 and CD103. Approximately 94 - 98% of the so-defined Treg cells expressed Foxp3 (data not shown).

The flow cytometry pointed out, that the proportion of Treg cells was significantly reduced in OVA (4.2 %) - and OVA/DHFR (4.5 %) - controls compared to naïve MLNCs (5.4 %) (p < 0.02). Mesenteric lymphocytes of OVA/Av17-treated animals showed a slight increase (6.3%) as compared to the naïve group but this ascent was not significant. However, Treg cells in control versus Av17-treated animals demonstrated a significant alteration. (p < 0.007, Av17 to OVA; p < 0.04, Av17 to DHFR; Fig. 33A).

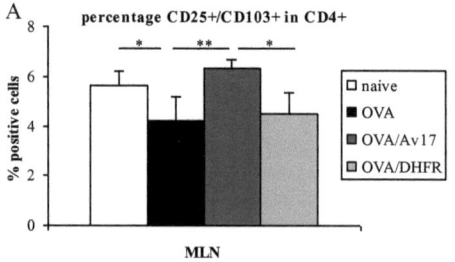
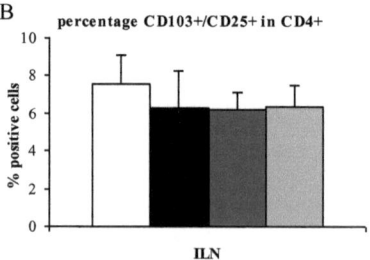

FIGURE 33: Percentages of CD4/CD25/CD103 positive cells in (A) mesenteric lymph nodes (MLN) and (B) inguinal lymph nodes (ILN). Shown are mean values of groups. naive: i.p. and e.c. PBS-treated mice; OVA: i.p. and e.c. ovalbumin-treated mice; OVA/Av17: i.p. and e.c. OVA-treated mice injected with filarial cystatin;

OVA/DHFR: i.p. and e.c. OVA-treated mice injected with control protein DHFR. Representative data of two experiments with 3-6 animals. *p < 0.05, **p < 0.005.

Analysis of inguinal lymph nodes revealed unaltered proportions of Treg cells for all tested groups (about 6.5 %, Fig. 33B), pointing toward only minor or no involvement of Treg cells in cystatin effects on dermatitis as Tregs in mesenteric lymph nodes should not be directly involved in amelioration of skin lesions.

3.3. *Heligmosomoides polygyrus* infection and allergy

3.3.1. Infection with the gastrointestinal nematode *H. polygyrus* ameliorates allergic airway inflammation but not the development of atopic dermatitis

Many reports on worm infections and allergy describe a strong negative correlation, resulting in less allergies in worm-infected individuals (see introduction). However some studies also point out positive correlations and enhancement of allergies by parasites. Apart from "the state" of infection (chronicity vs acute, as well as adult worm numbers) and the genus of the parasite (Smits 2007, Boitelle 2005) there also seem to be differences depending on the quality of the allergic disease and the biology or location of the parasite. To clarify the role of a gastrointestinal nematode infection (*H. polygyrus*) on allergen mediated sensitization and allergic disease two murine models were studied: ovalbumin (OVA)-induced airway disease, associated with the mucosal compartment and OVA-induced skin inflammation, linked to cutaneous tissue.

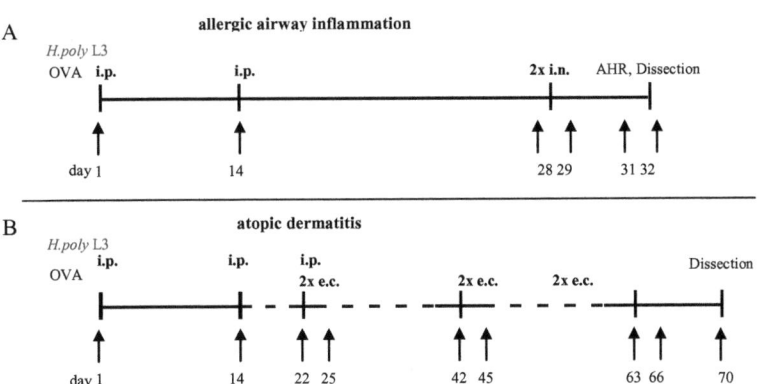

FIGURE 34: Model of worm infection in (A) allergic airway inflammation and (B) atopic dermatitis. Mice were sensitized intraperitoneally with the allergen ovalbumin (OVA i.p.) and challenged intranasally (i.n.) in the asthma model and epicutaneously (e.c.) in the dermatitis model. Infection with the gastrointestinal nematode on day 1 with infective larval stage 3 (*H.poly* L3).

In both approaches, BALB/c mice were infected orally with 150 L3 of *H. polygyrus* on the first day of intraperitoneal sensitization with the model allergen ovalbumin (OVA). Subsequently, mice were challenged with OVA via the airways (i.n.) (Fig. 34A) or via OVA-saturated skin patches (e.c.) (Fig. 34B).

In the asthma model both the OVA-group and *H. polygyrus*-infected mice that were sensitized and challenged with OVA (OVA/Hp-group), showed significantly increased total cell numbers in BALF ($p < 0.009$, $p < 0.01$) in comparison to naïve control mice (Fig. 35A). However, clear differences in the type of recruited cells were found between the OVA-group and the OVA/Hp-group. In the OVA-group the increase in total cells was due to a significant increase of eosinophils ($p < 0.009$). In contrast, in OVA/Hp-animals eosinophils in the BALF were significantly decreased ($p < 0.03$) but a trend towards an increase in macrophage recruitment into the lung ($p < 0.06$; Fig. 35A) was observed. Helminth-infected control groups, which were neither sensitized nor challenged with allergen, showed no increase of BAL cell numbers (data not shown).

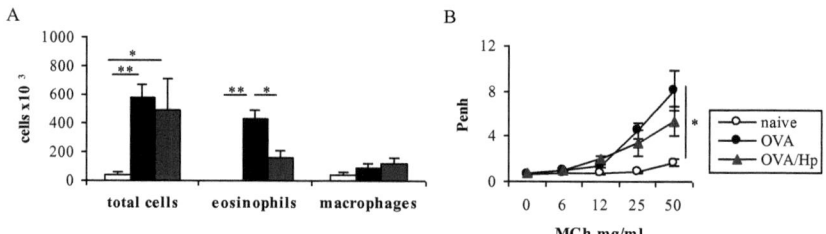

FIGURE 35: Infection with the gastrointestinal nematode *Heligmosomoides polygyrus* alters (A) eosinophil and macrophage numbers in BALF and (B) airway hyperreactivity of infected animals in response to increasing metacholin (MCh) concentrations in the asthma model. Shown are mean values of cell numbers and Penh (pause enhanced) values. naïve: PBS-treated animals; OVA: ovalbumin (OVA)-treated animals; OVA/Hp: OVA-treated and *H.polygyrus* infected mice. Representative data of 2 individual experiments with 5-6 animals per group *$p < 0.05$; **$p < 0.005$.

Measurement of airway hyperreactivity (AHR) in response to metacholin (MCh) stimulation of OVA/Hp-mice revealed a trend towards a down-regulation, although these changes did not reach statistical significance in comparison to the OVA-group (Fig. 35B).

These data show that the infection with *H. polygyrus* has a positive influence on airway inflammation, clearly reflected by eosinophil numbers, and thereby interferes with the development of airway disease.

These findings were in contrast to the influence of the intestinal worm infection on the non-mucosal allergic reaction, the murine model of OVA-induced atopic dermatitis.

After induction of eczema by systemic (i.p.) sensitization with ovalbumin and subsequent epicutaneous (e.c.) challenge with the allergen the clinical skin score was assessed. Interestingly, a concomitant infection with *H. polygyrus* did not improve clinical skin score of eczematous mice (Fig. 36A) but rather aggravated lesions (mean skin scores 7.4 in OVA-group, 8.4 in OVA/Hp-group, p < 0.02). These findings were in accordance with an increase

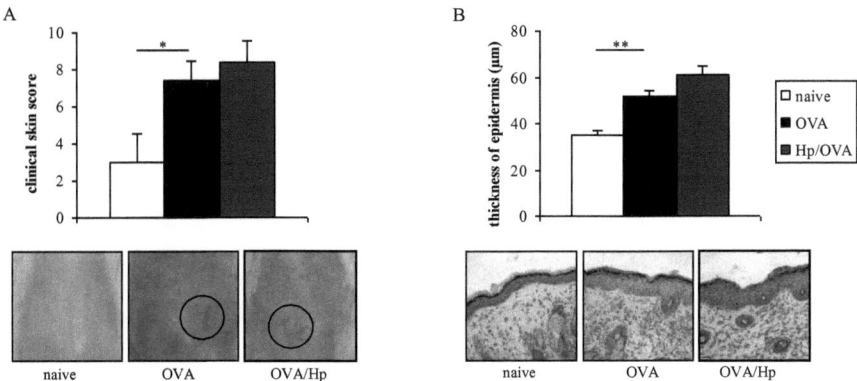

in the thickness of epidermis observed in histological sections of the eczema region. Thickness of epidermis was not significantly different between the OVA- and the OVA/Hp-group (52µm and 61µm; p < 0.001; Fig. 36B) but again a trend towards an aggravation by the nematode infection was detected.

FIGURE 36: Infection with *H. polygrus* in the dermatitis model aggravates (A) clinical skin score and (B) increase of thickness of epidermis. In the top panel mean values of groups are shown, the base panel consists of representative pictures of eczema (patch region indicated by a black circle) and epidermis sections. Measurements were performed by A. Dahten. naïve: PBS-treated animals; OVA: ovalbumin (OVA)-treated animals; OVA/Hp: OVA-treated and *H. polygyrus* infected mice. Representative data of 2 individual experiments with 5-6 animals per group. *p < 0.05; **p < 0.005.

These data indicate that *H. polygyrus* infection rather increased than ameliorated atopic dermatitis which stands in contrast to the ameliorative effect in the asthma model.

3.3.2. Allergen-specific humoral immune response of worm-infected mice in the asthma and dermatitis model is altered

Analyzing allergen-specific and total IgE in sera of mice in the asthma model clearly showed decreased levels of allergen-specific IgE (p < 0.002) in OVA/Hp-mice. In contrast, total IgE levels were significantly increased (p < 0.002) in the OVA/Hp group (Fig. 37A, B). Furthermore, allergen-specific IgG1 (p < 0.02) in sera of OVA/Hp-treated mice was significantly decreased (Fig. 37C). Allergen-specific IgG2a levels remained unaltered in the helminth-infected mice (data not shown).

Analysis of IgE in sera of A.D.-mice showed that concomitant infection with *H. polygyrus* significantly reduced allergen-specific IgE (p < 0.01, Fig. 37D), whereas total IgE was significantly elevated as compared to the OVA-group (Fig. 37E). Furthermore OVA-specific IgG1 was clearly down-modulated by worm infection (Fig. 37F), reflecting the observations of the asthma model.

Allergen-specific IgE was additionally determined by a basophil release assay (RBL-assay). No changes in mediator release were detected when comparing the OVA-group with the OVA/Hp-group in both models (not shown).

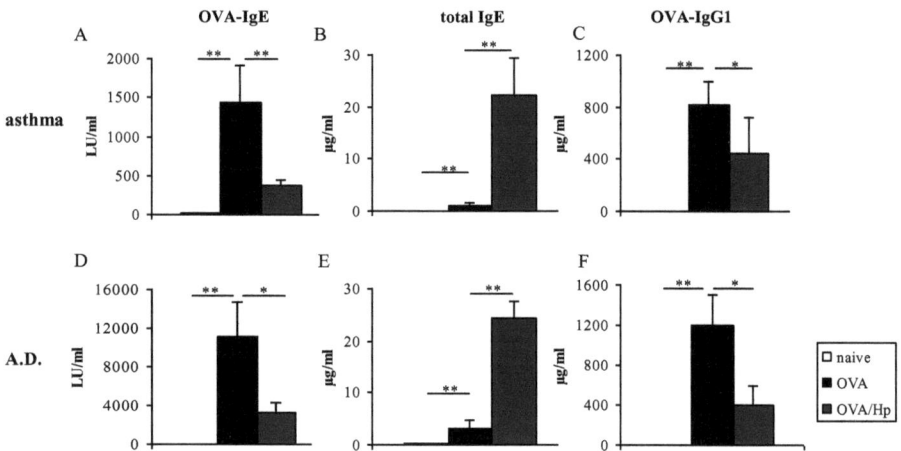

FIGURE 37: Antibody levels in sera of animals in the asthma model (upper panel) and the atopic dermatitis (A.D.) model (lower panel). Shown are mean values of (A, D) ovalbumin (OVA)-specific IgE, (B, E) total IgE and (C, F) OVA-specific IgG1. LU indicates lab units. naïve: PBS-treated animals; OVA: ovalbumin (OVA)-treated animals; OVA/Hp: OVA-treated and *H. polygyrus* infected mice. Representative data of 2 individual experiments with 5-6 animals per group. *p < 0.05; **p < 0.005.

These data indicate that in both allergy models a concomitant infection with the gastrointestinal nematode led to significantly diminished allergen-specific antibody concentrations in sera whereas total IgE levels remained elevated.

3.3.3. Local and systemic cytokine analysis in asthma and dermatitis differ

To determine the mechanism by which the nematode-infection induced amelioration of the allergic airway inflammation in comparison to pathological skin responses in the A.D.-model systemic and local cytokine production was analyzed. Stimulation of splenocytes with the allergen OVA led to reduced levels of IL-4 (p < 0.06, Fig. 38A) and IL-10 (p < 0.002, Fig. 38C) in the OVA/Hp-group in comparison to the OVA-group in the asthma model. No differences were found with regard to IL-5 production (Fig. 38B). In contrast, stimulation of spleen cells of OVA/Hp animals with parasite antigen led to secretion of high amounts of IL-4, IL-5 (p < 0.008) and IL-10 (p < 0.002) (Fig. 38G, H, I), which could not be observed in

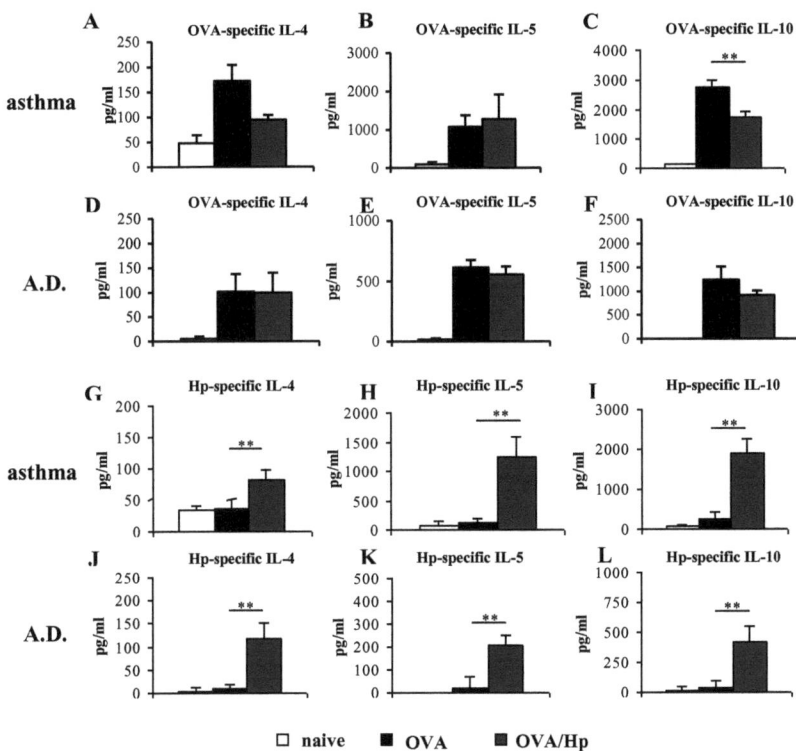

uninfected animals.

FIGURE 38: Cytokine production of splenocytes in the asthma and the dermatitis (A.D.) model. Ovalbumin (OVA)-specific production of (A) IL-4, (B) IL-5 and (C) IL-10 in the airway inflammation model. OVA-specific secretion of the cytokines (D) IL-4, (E) IL-5 and (F) IL-10 in the dermatitis model. Cytokine secretion upon stimulation of splenocytes with *H.polygyrus* antigen (Hp) in the asthma model: (G) IL-4, (H) IL-5 and (I) IL-10. Hp-specific cytokine production the in dermatitis model: (J) IL-4, (K) IL-5, (L) IL-10. naïve: PBS-treated animals; OVA: ovalbumin (OVA)-treated animals; OVA/Hp: OVA-treated and *H. polygyrus* infected mice. Representative data of 2 individual experiments with 5-6 animals per group. *p < 0.05; **p < 0.005.

In comparison, allergen-specific cytokine production in spleen culture supernatants of A.D.-animals showed no significant differences between the OVA- and OVA/Hp-mice regarding the cytokines IL-4, IL-5, IL-10 (Fig. 38D, E, F). However, stimulation of spleen cultures with parasite antigen resulted in up-regulation of IL-4 (p < 0.0019), IL-5 (p < 0.004) and IL-10 (p < 0.0098) (Fig. 38J, K, L) in the OVA/Hp group, as observed in the asthma model.

Results

Interestingly, Hp-specific IL-10 production in the A.D.-model was profoundly lower compared to worm-infected animals in the asthma model.

To investigate cytokines at the site of inflammation BAL fluid was analysed for production of IL-4, IL-5, IL-10 and TGF-β. Only TGF-β measurement in BAL fluid exhibited differences between the OVA-group and the OVA/Hp-group, although the decrease of TGF-ß production

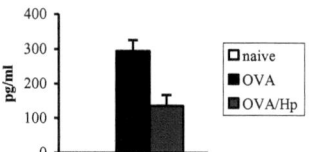

FIGURE 39: Measurement of transforming growth factor-ß (TGF-ß) in bronchoalveolar lavage fluid (BALF) of mice in the asthma model. naïve: PBS-treated animals; OVA: ovalbumin (OVA)-treated animals; OVA/Hp: OVA-treated and *H. polygyrus* infected mice. Representative data of 2 individual experiments with 5-6 animals per group.

in the OVA/Hp group was not statistical significant (Fig. 39).

In the dermatitis model cytokine mRNA was measured in the inflamed skin via real time PCR. The analyses indicated a trend towards up-regulated IL-4- and IL-10- mRNA levels though this did not reach statistical significance. However, TGF-β mRNA amounts in the OVA-Hp-group were significantly reduced compared to naïve animals ($p < 0.042$) (Tab. 3).

compared to OVA	fold change (mean)		
	OVA	naïve	OVA/Hp
IL-4	1.0 +/- 0.15	0.5 +/- 0.01	2.14 +/- 0.76
IL-10	1.1 +/- 0.22	0.56 +/- 0.19	2.10 +/- 0.70
TGF-ß	1.7 +/- 1.00	4.3 +/- 2.30	0.42 +/- 0.13

TABLE 3: Skin patch regions were analysed in real time PCR for mRNA of cytokines. Shown are fold changes in local cytokines compared to OVA-group indicated by mean values +/- SEM. Green: down-regulation of mRNA levels; Red: up-regulation of mRNA levels; OVA: i.p. and e.c. ovalbumin-treated mice; naïve: i.p. and e.c. PBS-treated mice; OVA/Hp: i.p. and e.c. OVA-treated mice infected with *H. polygyrus*.

Hence, the nematode infection exhibited a different impact on the cytokine production within the two allergy models. So far, it seems that lower amounts of Hp-specific IL-10 or lack of TGF-ß in lesions could be the reason for the detrimental effect of *H. polygyrus* in A.D.

3.3.4. Cellular infiltration in atopic skin is partly altered by *H.polygyrus* infection

Although the clinical outcome was aggravated and the thickness of epidermis was elevated in OVA/Hp A.D.-mice, differences in the numbers of cells infiltrating the skin lesions were observed. Skin sections stained for $CD4^+$ T cells showed significantly increased infiltration of T helper cells in OVA-mice mice suffering from dermatitis in comparison to naïve mice ($p < 0.001$). In contrast, numbers of infiltrating T cells were significantly decreased in OVA/Hp- mice ($p < 0.001$, Fig. 40A). However, infiltration of mast cells was found to be significantly enhanced in OVA/Hp-mice in inflamed skin areas (p<0.007 compared to OVA,

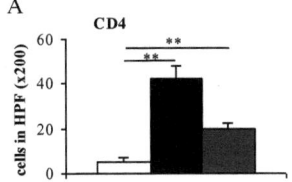
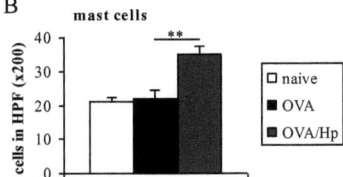

Fig. 40B).

FIGURE 40: Cells infiltration in patched skin regions in the dermatitis model. (A) T helper cells (CD4) and (B) mast cells were stained in skin sections. Shown are mean values of cell numbers counted in high power fields (HPF) in a 200 fold magnification. naïve: PBS-treated animals; OVA: ovalbumin (OVA)-treated animals; OVA/Hp: OVA-treated and *H. polygyrus* infected mice. Staining was performed by A. Dahten. Representative data of 2 individual experiments with 5-6 animals per group. **p < 0.005.

These data indicate, that *H. polygyrus* infection has no positive influence on the clinical outcome of atopic dermatitis, albeit it induces significant changes in cell recruitment into the inflamed skin. Although not described in this dermatitis model so far, it is tempting to speculate that the increase of mast cells found in skin of worm infected mice might account for aggravation of inflammation, mediated via a different mechanism.

3.3.5. *H. polygyrus* infection leads to induction of regulatory T cells in mesenteric and peribronchial but not in inguinal lymph nodes

To gather information whether regulatory T cells are involved in the influence of *H. polygyrus* on these two atopic immune responses Treg cells, characterized by the expression of the surface markers $CD4^+CD25^+CD103^+$, in the corresponding lymph nodes were analyzed. In the asthma model analysis of peribronchial lymph node cells (PBLNC) of OVA/Hp-mice showed a significant increase in Treg cells (p < 0.05) compared to the OVA-group (Fig. 41A, B). More than 95 % of the Treg cells were determined to be Foxp3 positive (data not shown). Such an increase in Treg cells was not found in PBLN cells of *H. polygyrus*-infected mice that were not sensitized and challenged with the allergen (data not shown). A significant increase of Treg cells could also be found in the mesenteric lymph node cells in the OVA/*Hp* group of the asthma model (p < 0.05; Fig. 41A, C).

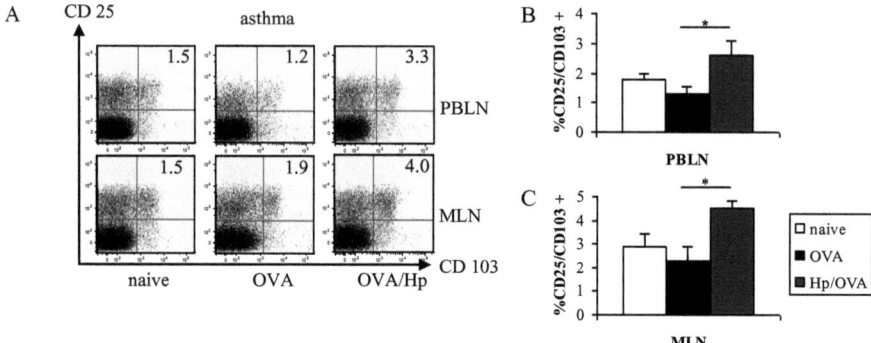

FIGURE 41: Analyses of T regulatory cells characterized by the expression of the surface markers CD4, CD25, CD103. (A) Representative dot plots of peribronchial lymph node cells (PBLN) and mesenteric lymph node cells (MLN) in animals of the asthma model. (B) Mean values of Treg numbers in PBLN of groups. (C) Mean values of Treg numbers in MLN in groups. naïve: PBS-treated animals; OVA: ovalbumin (OVA)-treated animals; OVA/Hp: OVA-treated and *H. polygyrus* infected mice. Representative data of 2 individual experiments with 5-6 animals per group. *p < 0.05.

Regarding the numbers of $CD4^+CD25^+CD103^+$ regulatory T cells in the OVA/Hp-group of the A.D. model, a significant increase of Treg numbers in mesenteric lymph node cells was determined (p < 0.05; 4 % to 7 %; Fig. 42A, C). However, this significant increase in MLNC was in contrast to the situation found in the skin-draining inguinal lymph node cells. There, no altered Treg numbers were observed in the OVA/Hp-group (6.2 % to 6.8 %; Fig. 42A, B).

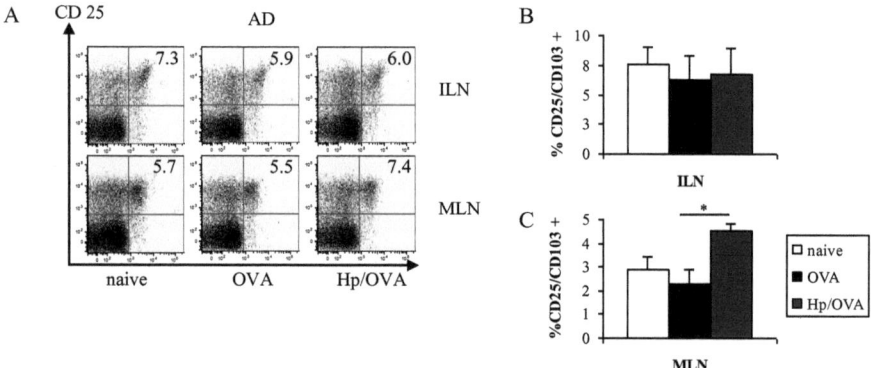

FIGURE 42: Analyses of T regulatory cells characterized by the expression of the surface markers CD4, CD25, CD103. (A) Representative dot plots of inguinal lymph nodes (ILN) and mesenteric lymph nodes (MLN) in animals of the dermatitis (A.D.) model. (B) Mean values of Treg numbers in ILN of groups. (C) Mean values of Treg numbers in MLN of groups. naïve: PBS-treated animals; OVA: ovalbumin (OVA)-treated animals; OVA/Hp: OVA-treated and *H.polygyrus* infected mice. Representative data of 2 individual experiments with 5-6 animals per group. *p < 0.05.

Furthermore, Foxp3 in skin lesions was measured by real time PCR. In skin lesions of the OVA/Hp-group no Foxp3 signal could be determined. Strikingly, also in atopic animals without worm infection Foxp3 in skin lesions was dramatically decreased (Tab. 4).

	fold change		
compared to OVA	**OVA**	**naive**	**OVA/Hp**
Foxp3	1.003 +/- 0.2	10.00 +/- 5.2	not detectable

TABLE 4: Skin patch regions were analysed via real time PCR for mRNA levels of the regulatory T cell marker Foxp3. Shown is the fold change of mRNA amounts compared to the OVA-group indicated by mean values +/- SEM. OVA: i.p. and e.c. ovalbumin-treated mice; naïve: i.p. and e.c. PBS-treated mice; OVA/Hp: i.p. and e.c. OVA-treated mice infected with *H. polygyrus*.

These data point in the direction that *H. polygyrus* infection leads to protection in airway inflammation because of an elevation in Treg numbers in the draining peribronchial lymph nodes (PBLN). Regarding atopic dermatitis, the data suggest that the gastrointestinal nematode is not able to enhance Treg numbers in the draining inguinal lymph nodes (ILN) or in patched skin regions, thereby having no ameliorative effect on this cutaneous model.

4. Discussion

4.1. Av17 and Allergic Airway Hyperreactivity (AHR)

4.1.1. Cystatin treatment modulates the allergic Th2 response

Infections with parasitic worms bear the potential to suppress allergic and inflammatory immune responses. Negative correlations between helminth infection and allergic disease have been observed in human field studies and especially in animal models (Maizels 2004, Fallon 2007). This work aimed to study if such protective effects could be exerted by single immunomodulatory proteins and provides evidence that the cysteine proteinase inhibitor of the parasitic nematode *Acanthocheilonema viteae* (Av17, cystatin) can indeed suppress allergic inflammation.

The *in vivo* experiments with recombinant filarial cystatin, expressed in *E. coli*, clearly show that application of this helminth immunomodulator in a murine model of ovalbumin-induced airway inflammation and hyperreactivity, during or even after allergen sensitization, counteracted the Th2 effector mechanisms responsible for allergic airway disease. The downmodulation of the Th2 response was accomplished in several ways.

First, cystatin treatment significantly reduced the recruitment of inflammatory cells, particularly eosinophils, into the lungs. Blood and tissue eosinophilia are hallmark signs of helminth infection, allergy and asthma (Prussin 2006). The release of effector molecules by these cells leads to tissue damage (Trivedi 2007, Rothenberg 2006). This mechanism was effectively prevented by interference with eosinophil recruitment by cystatin treatment.

Second, after application of cystatin the production of allergen-specific and total IgE was strongly downmodulated. This resulted in less efficient sensitization of mast cells and basophils (Bradding 2006), as determined by decreased degranulation of basophils sensitized with sera of cystatin/OVA-treated mice. Histamine, released by mast cells, has important immunomodulatory functions, as application of Histamine-1-receptor (H1R)-antagonists prior or during sensitization prevents allergen induced airway inflammation and AHR by altering the expression of costimulatory molecules on APCs (Blümchen 2004).

Another mechanism influencing mediator release might be coaggregation of FcεRI and FcγRIIB that activates an ITIM (immunoreceptor tyrosine-based inhibition motif) that prevents activation of mast cells (Kraft 2006). For this coaggregation cystatin would have to act as a dimer or even polymer *in vivo*. These structures were observed in preparations of chicken and human cystatin (He 2005; Janowski 2001), and have also been found for Av17 (T. Buhrke, personal communication).

Third, cystatin treatment reduced local production of the Th2 cytokine IL-4 in BALF as well as systemic IL-4 production by spleen cells. IL-4 plays a major role in the development of allergic reactions: It is crucial in the induction of the production of IgE by mediating class-switch (Poulsen 2007). Furthermore IL-4 promotes Th2 responses by stimulating Th2 cells and directing naïve T helper cells into Th2 direction (indirect effect via dendritic cells, Spellberg 2001). Interleukin-4 is also described to enhance tissue homing of inflammatory effector cells (Romagnani 2004). Therefore, it is assumed that the reduction of IL-4 levels by cystatin treatment contributed to the inhibition of IgE production and the lowering of Th2 responses in this model.

Absence of high IL-4 levels could also provide the possibility to switch the immune response from "atopic" Th2 to "non-atopic" Th1. This aspect has been proven to be important in the context of bacterial immunomodulators, as TLR2 and TLR4 agonists (Velasco 2005, Gerhold 2006). These components act through stimulation of Th1 responses, as evidenced by elevated production of IFN-γ and other Th1 cytokines, thereby blocking Th2 response. Th1 and Th2 cells cross-regulate each other by expression of IFN-γ to stop IL-4 secretion and differentiation of Th0 cells to Th2 cells or by IL-4 and IL-10 secretion that blocks IL-12 and IFN-γ production and therefore differentiation to Th1 respectively (Gajewski 1988, Ohmori 1997, Ito 1999).

However, filarial cystatin stands in contrast to typical TLR ligands: There was never production of the cytokines IFN-γ or IL-12 observed in cystatin-treated animals, arguing for a different mechanism. Furthermore filarial cystatin was shown to be able to inhibit Th2-related as well as macrophage-related inflammation (colitis, Schnoeller 2008) and, as presented in this work, also mixed Th1/Th2 responses (dermatitis), which is clearly not possible by the Th1/Th2 shift described for TLR2 and TLR4 agonists. Regarding possible contaminations of the protein preparation, it has to be said, that the classical TLR4 ligand LPS was described to ameliorate airway eosinophilia and OVA-IgE only in high concentrations (about 10 µg per mouse) and had no effect on AHR development (Gerhold 2002). Low dose administration of LPS (<10 ng; i.p.) was shown to even enhance airway eosinophilia (Delayre-Orthez 2004). The low dose applications described in literature still comprise more endotoxin than leftover LPS found in cystatin preparations (max. 40 pg per application). Also positive effects of BCG (Bacillus Calmette Guerin, *Mycobacterium bovis*) are thought to be mediated by a switch towards Th1 response, involvement of TLR receptors and strong IL-12 production (Barlan 2006). However IL-12 was also not induced by cystatin treatment.

4.1.2. Role of the anti-inflammatory cytokines IL-10 and TGF-ß

Filarial cystatin has been described in recent publications as a potent immunomodulator *in vitro*. In these studies Av17 (Hartmann 1997) and its human counterpart Ov17 (*Onchocerca volvolus* cystatin) (Schoenemeyer 2001) could be shown to induce IL-10 secretion in splenocytes or PMBC respectively. After depletion of monocytes, IL-10 secretion was abrogated, pointing to macrophages as the main producing cells. In addition parasitic cystatin was proven to inhibit proteinases (human cathepsins L and S) and several costimulatory molecules (CD86, HLA-DR) involved in antigen-processing and presentation. Further studies pointed out, that these effects were characteristic of parasite-derived cystatins as cystatin of the soil-nematode *C. elegans* (Cysele 2) did not lead to IL-10 induction and inhibited different cathepsins (Schierack 2003). Finally Av17 could be shown to be a potent suppressor of T cell proliferation probably mediated via IL-10.

Keeping this in mind brings the anti-inflammatory cytokine IL-10 in the spot light. To corroborate this hypothesis, in this work, application of anti-IL-10-receptor (αIL10R) antibodies to block the mediator was performed. Blocking of IL-10 receptor completely reversed the effect of cystatin on cell recruitment, production of IgE and airway hyperreactivity.

Tournoy and co-workers described that IL-10 therapy in mice successfully inhibited airway inflammation and nonspecific airway responsiveness (Tournoy 2000). Furthermore IL-10 gene delivery in mice suppresses airway hyperreactivity (AHR) by down-modulation of APC functions (Nakagome 2005). Taken together with reports suggesting that production of IL-10 by Treg cells plays a central role in control of allergic airway disease (Hawrylowicz 2005b) and that systemic production of IL-10 is capable to inhibit the development of airway hyper-responsiveness and allergic inflammation (Fu 2006, Urry 2006), it seems fitting to speculate that IL-10 is a key element in the cystatin-induced immunomodulation in asthma. Even more intriguing, the depletion of macrophages led to an abrogation of IL-10 production and aggravation of inflammation, whereas Treg depletion did not influence the systemic release of IL-10 and did not block the protective cystatin effect to the same extend.

IL-10 was reported to modulate Th2 responses by a broad range of suppressive mechanisms: By suppression of allergen-specific IgE production and concomitant induction of non-inflammatory antibody isotypes, by reduction of pro-inflammatory cytokines released by mast cells, basophils and eosinophils, and by indirect interference with Th2-associated phenomena such as mucus production (Taylor A 2006, Larche 2006). Thus, IL-10 seems indeed to be capable of redirecting pathologic allergic responses. There is also evidence for positive effects of IL-10 from human studies: in allergic patients undergoing successful immunotherapy IL-10 levels are increased (Akdis 2006, Bohle 2007). IL-10 has also been shown to inhibit IgE induced mast cell activation in humans (Royer 2001).

In addition to its effects on Th2 responses IL-10 is also known to modulate the maturation of dendritic cells and inhibit the expression of MHC II and co-stimulatory molecules (Moore 2001).

As Av17 is derived from a parasite it is also fitting to find importance of IL-10 in its protection against asthma, as this is in accordance with reports on worm infection and asthma. IL-10 was described to modulate allergic immune responses in mice infected with the gut nematode *Heligmosomoides polygyrus* or the blood fluke *Schistosoma mansoni* (Kitagaki 2006, Mangan 2004) shown by the use of IL-10 knockout mice or αIL-10R, respectively. Also human studies point in this direction: The low response to the house dust mite (HDM) allergen Derp1 in skin prick tests of worm infected children in Gaboon could be linked to increased IL-10 production and no association between polyclonal IgE antibodies and skin-test results was found (van den Biggelaar 2001). IL-10 is also crucial in mediating protection to food allergy by *H. polygyrus* as Bashir (2002) and co-workers could demonstrate by the use of αIL-10R antibodies.

Of interest, some other studies demonstrate that helminth-induced protection against allergic disease can be independent on IL-10 (Trujillo-Vargas 2007) and propose a role for TGF-ß (Wilson 2005). TGF-ß was described in recent studies to have very distinct effects on asthma-like disease. Application of antibody against TGF-ß in a mouse model resulted in suppression of pulmonary fibrosis but enhancement of OVA-induced AHR and had no significant effect on airway inflammation and eosinophilia (Alcorn 2007). Fattouh and co-workers monitored effects in a model of HDM-sensitization. They could induce airway inflammation and remodelling in mice treated with αTGF-ß antibody and furthermore eosinophilic infiltrate was exacerbated and led to increased airway hyperreactivity (Fattouh 2008). TGF-ß is described to play an important role in the differentiation of Th17 cells. This might explain why TGF-ß sometimes revealed proinflammatory effects (Veldhoen 2006).

In this work, transforming growth factor TGF-ß was analysed by two different techniques, ELISA and real-time-PCR. However, following cystatin treatment there was never any increase detected, but a tendency towards decreased TGF-ß mRNA amounts in lungs. Interestingly, in human studies, asthmatic patients provide higher levels of TGF-ß mRNA, reflecting fibrosis (Vignola 1997, Kenyon 2003). Hence, a decrease of TGF-ß in cystatin-treated animals might reflect lowered fibrosis. Still, systemic induction of TGF-ß might bear the potential to suppress allergic airway inflammation, as it has been described as a potent anti-inflammatory protein (Veldhoen 2006).

4.1.3. Involvement of macrophages

Thinking about a cell producing the mediator(s) and being the target of cystatin, the data obtained and recent studies in our lab (Hartmann 2003, Schoenemeyer 2001) speak for a crucial role of macrophages in the protective effect of filarial cystatin in the murine asthma model. Depletion of macrophages with clodronate liposomes two days prior to airway allergen challenge was found to completely ablate the anti-allergic effect of cystatin, including AHR, eosinophil infiltration, OVA-IgE and partly IL-4.

Liposomes are artificially prepared lipid vesicles that can be used to encapsulate strongly hydrophilic molecules such as clodronate, solved in aqueous solutions. Clodronate (dichloromethylene-bisphosphonate, Cl_2MBP) is a bisphosphonate that is usually prescribed as a bone resorption inhibitor and antihypercalcemic agent (en.wikipedia.org/wiki/ clodronate). Freely solved clodronate will not cross liposomal or cellular phospholipid membranes but after injection, liposomes will be ingested and digested by macrophages followed by intracellular release and accumulation of clodronate. At a certain intracellular concentration, clodronate induces apoptosis of the macrophage (www. clodronateliposomes. org). It is noteworthy to say that treatment with clodronate liposomes (MLV) mainly targets macrophages (because of the lipids used for preparing MLV) but also a small part of DC may be killed. However, all other potentially IL-10 producing cells (B cells, T cells, NK cells etc.) remain present (Van Rooijen 1994; Anthony 2006).

In addition to abrogation of cystatin`s protective effects, macrophage depletion led to significantly diminished IL-10 levels in splenocytes of cystatin-treated animals, speaking for the macrophage as the main producer. Strikingly the major source of IL-10 *in vivo* (in humans) is the macrophage (Asadullah 2003).

Macrophages have recently been of intense interest: They are important in the initiation of inflammation but also in the resolution of this response: Macrophages are necessary for the clearance of granulocytes and the elaboration of anti-inflammatory mediators contributes to dissolution of the inflammatory response (Zhang 2008). Macrophages can produce TGF-ß and thereby contribute to wound healing. Furthermore, they can produce IL-10 that may inhibit the synthesis and activation of many inflammatory cytokines (Mege 2006, O´Garra 2007). In addition, macrophages produce inflammatory tumor necrosis factor (TNF) that resolves inflammation by binding to death-domain containing TNF-R1 and induces cellular apoptosis

(Shen 2006). Today macrophages are broadly classified in two main groups, which are described according to static functional and phenotypic criteria (Mantovani 2005; Fig. 43). Classically activated macrophages (M1) are typically activated by IFN-γ and LPS or other TLR ligands leading to TNF secretion. They usually produce high levels of nitric oxide (NO) and exhibit low arginase activity but high CD86 expression. Classically activated macrophages are "effective APCs", meaning that they stimulate T cell proliferation. M1 are also characterized by production of IL-12 and give rise to Th1 cells (Martinez 2008).

Alternatively activated macrophages (M2) are further divided into M2a, M2b and M2c. When talking about alternatively activated macrophages (AAM) mostly M2a cells are referred to. They originate after exposure to IL-4 or IL-13 and are characterized by various markers. High expression of surface IL-4 receptor α chain (IL-4Rα) and mannose receptor (CD206) as well as expression of the newly identified markers macrophage galactose-type C-type lectin expression (mMGL1 and mMGL2) and IL-27 receptor (WSX-1 = IL-27Rα) are characteristic (Edwards 2006, Rückerl 2006). Intracellular markers for M2a AAM are high expression of FIZZ1 (found in inflammatory zone 1, resistin-like molecule α), Ym1/2 (secretory lectin, chitinase family) and arginase 1. Arginase expression restricts availability of L-arginine, a substrate for iNOS. This prevents NO production by these cells and shifts arginine utilization to production of polyamines and proline which are needed in wound-healing (Zhang 2008). M2a macrophages fail to stimulate T cell proliferation and do not express CD86 on their surface. This group of AAM has been closely associated with parasitic disease (Hesse 2001) and may provide immunity during helminth infections (Brys 2005, Anthony 2007).

Macrophage characteristics can be influenced by glucocorticoides (IL-10 up), adenosine (TNF down, IL-10 up) and resolvins (derivatives of eicosapentaenoic acid, promote uptake of apoptotic cells) and the surrounding of IL-10 and TGF-ß. These macrophages are merged as M2b alternatively activated macrophages (Zhang 2008).

In close proximity to immune complexes in combination with IL-1ß or LPS (Toll-like receptor ligands) the third group of AAM, the M2c or type II macrophages emerge. Type II macrophages express high levels of NO and CD86, low arginase activity and stimulate T cell proliferation (give rise to IL-4 and IL-10 secreting Th2 cells). Interestingly they produce a lot of IL-10 but no IL-12 (Anderson 2002a, b, c) and are described as strong anti-inflammatory cells (Gerber 2001). Currently they are best characterized on mRNA level by the markers sphingosine kinase-1 (SPHK1 or SK-1) and LIGHT (TNF superfamily 14). Of interest, *in vitro* studies revealed that SPHK1 catalyzes the production of sphingosine-1 phosphate, which is necessary for C5a-triggered, intracellular Ca^{2+} signals (Melendez 2004), might play a role in retaining cell viability in endotoxin-stimulated MØ (Wu 2004) and controls overproduction of Th1 cytokines in T cells (Yang 2005). LIGHT can costimulate T cell responses (Tamada 2000) and transmit costimulatory signals into T cells thereby enhancing activation in response to suboptimal TCR interactions (Shi 2002, Wan 2002).

Type II macrophages have been suggested to play a role in exacerbation of visceral leishmaniasis where IgG-coated parasites can induce production of IL-10 from macrophages and allow disease progression (Miles 2005).

In general M2 macrophages promote Th2-associated effector functions and play a role in resolution of inflammation through endocytic clearance, trophic factor synthesis and reduced proinflammatory cytokine secretion (Martinez 2008). The new cytokine IL-21 seems to augment alternative macrophage activation and Th2 response as shown in experiments with IL-21R$^{-/-}$ mice, having an attenuated Th2 response to *N. brasiliensis*, *H. polygyrus* or *S. mansoni* (Pesce 2006, Frohlich 2007).

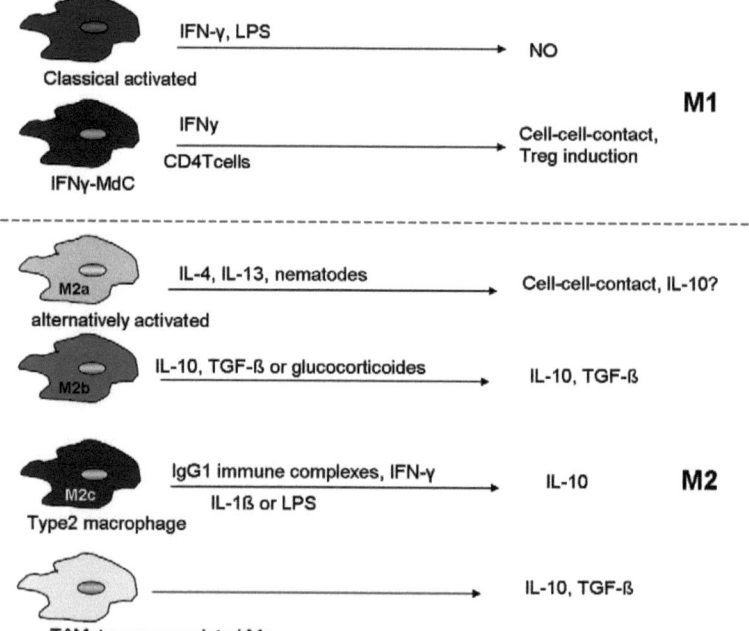

Figure 43: Overview of classically (M1) and alternatively (M2) activated macrophage populations. Stimuli to induce cells and main effector mechanisms of macrophages are indicated.

Helminth infections are associated with induction of alternatively activated macrophages (Herbert 2004, Gordon 2003, Hesse 2001, Loke 2000, Pesce 2006, Anthony 2007). In murine models of filarial infection (*Brugia malayi*) the nematode-elicited macrophages express phenotype characteristics of AAM and are potent suppressors of T cells proliferation (Nair 2005). Potent suppressive AAM also arise with *L. sigmodontis* infection and the immune suppression seems to be independent of IL-10 and CTLA-4 but partially dependant on TGF-ß (Taylor M 2006). Anthony and colleagues could show in a model of *H. polygyrus* infection that in secondary infection with the helminth, memory cells in the gut produce IL-4 and lead to the induction of AAM expressing IL-4Rα and CD206. The protective Th2 immune response is mediated by AAM which impair larval parasite health and mobility through an arginase-dependent mechanism (Anthony 2006). AAM can contribute to fibrosis and repair at the site of injury which might be of considerable importance during helminth infection but could also be extended to other "wounds" (Martin 2005).

However, also macrophages with suppressive activity that are not alternatively activated cells have been shown to be induced by injection of schistosome egg glycans. These $F4/80^+Gr1^+$ macrophages suppress T cell proliferation (Atochina 2001, Terrazas 2001). Furthermore schistosomes can modulate splenic macrophages to induce T cell anergy via a mechanism involving the costimulatory surface marker PD-L1 (Smith 2004).

Also macrophages isolated from tumors (Mantovani 2002) and glucocorticoid-treated macrophages (Frankenberger 2005) have been reported to preferentially secrete IL-10.

Another novel type of macrophage is defined by its state of activation. IFN-γ-stimulated monocyte derived cells (IFN- γ-MdC) differ in mode of generation, cell surface phenotype and function from classically activated macrophages. They arise only in the presence of CD40L-expressing CD4 T cells, M-CSF and IFN- γ. Interestingly, they enrich co-cultured T cell populations for $CD4^+CD25^+Foxp3^+$ Tregs by expansion of these cells and depletion of activated T cells. Furthermore, they express CD274 (PD-L1) but could not be shown to use this costimulatory pathway. Lymphocyte killing is cell contact and caspase dependent (Brem-Exner 2008) and transferred cells can protect from colitis.

Munn et al described a somewhat similar macrophage type derived of human PMBCs in presence of CD40L, IFN- γ and M-CSF. However, these classically activated macrophages influence T cell function through IDO (Indoleamine-pyrrole 2,3-dioxygenase) - dependent mechanisms (IDO-mediated tryptophan depletion) and lead to T cell apoptosis (Munn 1999, 1996).

Although definitions of different functional phenotypes of macrophages have scientific and diagnostic value it is quite possible that macrophages are not static and that functional and phenotypic changes may be just an evolving shift of functional activities in response to regulatory oscillations in their environment (Stout 2004).

Cystatin treated macrophages still produce NO *in vitro*, however *in vivo* application of cystatin did not lead to measurable amounts of IFN-γ or other Th1 cytokines *in situ* (in the lung) or systemically, speaking against an effect on/via classical activated macrophages.
In the setting of the analyzed asthma model the Th2 cytokines IL-4 and IL-13 can be found in mice suffering from asthma. Cystatin was found to mediate downmodulation of both these cytokines (especially IL-4) but nevertheless it is still feasible that in the beginning of treatment enough of these cytokines were present to alternatively activate macrophages. This would be in line with the fact that alternatively activated macrophages can develop after helminth infection (MacDonald 1998), are markedly cytostatic and prevent the proliferation of non-lymphoid as well as T cells. They have also been shown to reduce the proliferation of a Th2 cell clone *in vitro* (Allen 1998). AAM were found in the presence of filarial and other helminth infections (MacDonald 1998), where secretion of cystatin may also be found. Nonetheless, AAM have been described to have anti-inflammatory potential, for instance they are able to suppress T cell proliferation by secretion of the cytokines TGF-ß and IL-10 (Goerdt 1999, Reyes 2007). Interestingly, Hoves and co-workers could show alternatively activated macrophages to mediate anergy and induce regulatory T cells (Hoves 2006). AAM found in filarial infections were also described to use a different mechanism to interfere with T cell proliferation, namely comprising a receptor-mediated mechanism and cell-cell-contact (Loke 2000). Alternatively activated macrophages could indeed bear the potential to mediate protection of inflammation as found by application of cystatin.
Nevertheless in recent studies (T. Buhrke, unpublished observation), transcription profiling of murine macrophages after exposure to filarial cystatin revealed an up-regulation of IL-10 expression, but no changes in expression of markers of alternative macrophage activation (Arg, FIZZ, YM1).
Thus one of the other subpopulations of AAM might be generated in response to Av17 treatment. Cystatin-treated macrophages seem to resemble type II–activated macrophages (Edwards 2006) or M2c. As described above, type II macrophages are effective producers of IL-10, mRNA of the Th1 cytokine IL-12 is completely diminished but NO production is still active. Although characterization of all these possibilities in detail is done by co-workers and is not a part of this work, it is still intriguing to speculate about type II macrophage induction by cystatin. Cystatin can not only induce IgG1 antibodies to form immune complexes, it has

also been described to give a short stimulus of TNF-α before IL-10 production *in vitro* (Hartmann 2002). This could be enough stimulus to replace the usual costimulator LPS. Furthermore Av17 may also bind to TLR4, thereby providing the costimulus, maybe also by atypic binding, as Harnett (2006) could prove for another nematode immunomodulator, ES-62.

Macrophages have also been in the focus of other studies concentrating on asthma and / or helminth immunomodulation. Recently Niu and co-workers could unravel a novel pathway regulating inflammatory disease in the respiratory tract. They could show that airway inflammation with repeated exposure to inhaled antigen suppresses CD4 T cells and their effector response in the respiratory tract and that this suppression was not dependent on host lymphocytes, in contrast to Tregs described as mechanism for inhaled tolerance (Holt 1981, Akbari 2001, Ostroukhova 2004). Furthermore, suppression was antigen non-specific and associated with a marked expansion of TGF-ß1-expressing lung macrophages (Niu 2007). Smith and co-workers described, that infection with the trematode parasite *S. mansoni* prevents experimental colitis via a macrophage-mediated mechanism with strong infiltration of $F4/80^+CD11b^+CD11c^-$ macrophages into the colon lamina propria, which is transferrable to uninfected animals. So far, they excluded that the effect is mediated by alternatively activated macrophages via measurement of arginase in colon macrophages and use of $IL-4/IL-13^{-/-}$ that have defects in AAM (Smith 2007).

Taken together, it seems quite likely that cystatin treatment indeed leads to induction of IL-10-secreting type II macrophages that mediate protection of allergic airway inflammation.

4.1.4. Role of regulatory T cells (Tregs)

Another target cell of cystatin might be regulatory T cells. Tregs have been suggested to play a major role in several settings of protection against asthma (Hawrylowicz 2005a, Stock 2006) and especially in protection by parasitic worms (Kitagaki 2006, Wilson 2005). Leech and colleagues could show in a Derp1 model that transfer of $CD4^+CD25^+Foxp3^+$ Tregs led to resolution of allergic airway inflammation in an IL-10 independent way (Leech 2007). Of special interest, there have been human studies suggesting that allergy is due to a lack or functional insufficiency of these immune cells (Hartl 2007, Lin 2008).

In an animal model Wilson could transfer protection by MLNC transfer, but also by MLNC derived from IL-10 knockout mice. He found increased levels of IL-10 in C57BL/6 Derp1 mice and increased, but not significantly, TGF-ß in BALB/c OVA mice in BALF (Wilson 2005).

Discussion

Interestingly, the protection of asthma by transfer of OVA-stimulated CD4$^+$ T cells was only effective when cells were derived from Hp/OVA treated mice, not from Hp or OVA treated mice and dependent on IL-10 (Kitagaki 2006). Furthermore, T regulatory cells obtained from *O. volvolus* infected patients secrete IL-10 and TGF-ß and suppress proliferation (Satoguina 2002). This indicates controversial data on the involvement of IL-10 in Treg suppression in asthma.

Up to now, Treg cells have been roughly divided into natural Tregs (characterized by CD25, CD4 and Foxp3 expression), Tr1 cells characterized by secretion of unusually high levels of IL-10 and lower levels of TGF-β (Groux 1997, Buer 1998), and Th3 cells described to produce large amounts TGF-β and thought to be related with mucosal immunity and IgA production (Chen 1994, Fukaura 1996, Powrie 1996, Hafler 1997).

In this work a population of natural Treg cells was analyzed. The effector/memory-like regulatory T cell subset is characterized by the expression of CD25, CD4 and the integrin $\alpha_E\beta_7$ (CD103) and was shown to exhibit a particular high suppressive capacity (Lehmann 2002), especially in the context of *H. polygyrus* infection (Rausch 2008). In addition control co-stainings with Foxp3 (forkhead box transcription factor p3) ensured that Treg cells and not T effectors were analyzed.

Filarial cystatin was found to induce small but significantly elevated numbers of Treg cells in the lung draining lymph nodes, or in other words at the site of inflammation. Consequently, importance of this increase was analyzed by depletion of Treg cells. Treg cells were depleted with high efficiency by application of αCD25 antibodies in the asthma model 5 days before airway challenges. However, lack of Treg cells did not convert the protective effect of cystatin completely. Depletion of Tregs revealed only a trend of restoration of eosinophil numbers in the lung and no significant increase in AHR, arguing for only a partial involvement of Treg cell in Av17-protection.

However, one could argue that conversion of the protective cystatin effect was only partly because Treg cells were depleted by antibody application. Currently αCD25 application to deplete Treg cells is discussed controversly. Kohm et al suggest, that Treg cells are not depleted but CD25 expression is simply downmodulated by application of antibodies. His work refers mostly to the use of the depleting αCD25 antibody named 7D4 which has indeed produced bigger proportions of CD25 ^{low} cells but not a lack of them (Kohm 2006). However, Stephens and co-workers describe a significant effect of the depleting antibody PC61 as do Zelenay and Demengeot (Stephens 2006, Zelenay 2006). They could present convincing data that PC61 application does not only lead to functional inactivation but really lack of Treg cells. In the present work PC61 was used, speaking for efficient depletion of Treg cells as also observed in control FACS stainings.

Nevertheless, to be absolutely sure that no Treg cell is left, DEREG (depletion of regulatory T cell) mice would be the experiment of choice. These are chromosome-transgenic mice expressing a diphtheria toxin (DT) receptor-enhanced green fluorescent protein (GFP) fusion protein under the control of the foxp3 gene locus, allowing selective and efficient depletion of Foxp3$^+$ Treg cells by DT injection (Lahl 2007). However, it still remains to be established if our asthma model will work in these mice and if Treg cell can be depleted throughout the whole model.

Hence, cystatin has an effect on the recruitment of regulatory T cells no matter if they are crucial for protection or not. It is possible that changes occurring in other Av17-target cells, namely the macrophages, could create a milieu attracting Treg cell migration, proliferation or production. Furthermore, as our depletion experiments revealed that Treg cells contribute to the lowering of eosinophil numbers and OVA-specific IgE production, it might be that IL-10–independent Treg cell functions have a role in the anti-allergic effect of cystatin (Fehervari 2004, Miyara 2007).

4.1.5. Potential role of B cells

Interestingly, regarding other infections such as schistosomiasis or toxoplasmosis, B cells or Th1 cells are shown to be the relevant IL-10 sources and mediate suppression of anaphylaxis (Mangan 2004) or host protection (Jankovic 2007). A number of studies has been performed in the past, directing to an important role for regulatory B cells (Breg) in anaphylaxis, EAE (experimental autoimmune encephalomyelitis) and other inflammations (Mauri 2008).

However, B cell depletion has to be done in a very careful approach, as there are reports that B cell depletion and B cell deficient mice would interfere with the model (Hamelmann 1997a, b). A possibility to analyze B cells would be the transfer of B cells derived from spleen, a method described by Smits (2007) in the context of transfer of schistosome-dependent protection of airway inflammation. Alternatively, intracellular IL-10 staining of B cells could be performed. However, in a recent report, B cell depletion shortly before challenge with αIgM was shown to be feasible, as Mangan (2006) could demonstrate influence of B cells in male schistosome-mediated protection of AHR and lung inflammation and described no influence per se on this model, leaving another option, although influence on sensitization might be quite drastic.

Discussion

For the future, it seems very sensible to perform experiments to address involvement of these cell populations in the context of filarial cystatin, but maybe better in disease models that do not directly depend on antibody production.

4.1.6. Possible receptors translating Av17 signals

As the results obtained with Av17 in the airway hyperreactivity model point to the macrophage as the main target cell of Av17 it is also necessary to speculate on possible receptors the proteinase inhibitor cystatin could address. There are several hints pointing to an involvement of innate immunity receptors such as scavenger or Toll-like receptors.

Scavenger receptor CD36 is one potential candidate. It is predominantly expressed on macrophages, platelets and endothelial cells (Adachi 2006). Of particular interest, CD36 has been described as the receptor for the cell adhesion protein PfEMP1 of the malaria parasite *Plasmodium falciparum* and binding of PfEMP1 to CD36 leads to the induction of IL-10 (Urban 2001) which would fit to effects seen with Av17. Furthermore CD36 is linked to uptake of apoptotic material and thereby modulating DCs to a "passive" or "alternative" activation state, which could influence inflammatory responses (Urban 2001). Another possibility might be TGF-ß receptor II. Human cystatin has been shown to physically interact with the TGF-ß receptor, thereby antagonizing TGF-ß signaling in cancer cells (Sokol 2004). As TGF-ß is found to be up-regulated in asthmatic lungs in human patients and downmodulated in murine lungs treated with Av17, one could hypothesize that cystatin competes with host TGF-ß and thereby reduces asthma pathology, in addition to IL-10 induction which would be required to block for example IgE production. Furthermore, other pattern recognition receptors (PRR) as TLRs (Toll-like receptors) might be addressed. Regarding helminth infections three distinct TLRs have been described to be addressed so far: TLR2 (schistosomal lyso-PS; filarial AvTropo), TLR3 (schistosome egg dsRNA) and TLR4 (filarial ES-62) (van der Kleij 2002, Aksoy 2005, Goodridge 2005a). Of interest, the PC containing secreted glycolipid ES-62 targets TLR4 in an atypic manner compared to the classical ligand LPS. This was shown by the use of HeJ mice in which LPS cannot bind or signal via TLR4 because of a point mutation in the receptor. Nevertheless ES-62 signaling is still functional (Harnett 2006). It is possible that cystatin might trigger a similar cascade. Another interesting aspect is the inducible T cell costimulatory receptor CD137 (4-1BB). This member of the TNFR-family normally addresses CD8 T cells, however experimental costimulation of CD137 (by agonistic antibodies) was shown to inhibit Th2-mediated AHR, eosinophil recruitment and allergen-specific antibody production. CD137 is expressed on activated T cells, natural killer cells, NKT cells and Tregs and its natural ligand CD137L

(4-1BBL) is found on APCs (B cells, macrophages and DCs). CD137 stimulation increases Treg numbers but the protective effect is not blocked by application of PC61 (Sun 2006). So far, it is up to speculation if cystatin treated macrophages increase CD137L expression or if cystatin can directly bind to CD137.

Finally, it would also be feasible that cystatin binds to the FcγR as a complex with IgG1 antibodies and thereby generates immunmodulatory IL-10 producing Type II macrophages. This is strongly suggested by *in vitro* data obtained by colleagues (C. Klotz, P. Burda) and discussed in detail above (see type two macrophages).

4.1.7. Proteinase inhibitor function of Av17 and role of DCs

A. viteae cystatin has been described to be a potent cysteine proteinase inhibitor (Hartmann 2003). Therefore one could also assume that the proteinase inhibitor function of cystatin is needed for its protective effect. Layton and colleagues observed therapeutic effects of the broad-spectrum cysteine proteinase inhibitor E64 on allergic lung inflammation. E64 had mainly effects on extracellular cathepsins and thereby influenced migration and proliferation of allergen specific T cells. Furthermore, it was postulated that E64 could block or inhibit cell-cell interactions, regulate cell adhesion molecules and cytokine, chemokine and death receptor ligands by blocking proteinases needed for the formation of these surface molecules. However, in contrast to data obtained with Av17 application, they did not show any influence on IgE levels, mucus production or airway hyperreactivity (Layton 2001). In addition, *in vitro* studies indicate, that IL-10 production after cystatin stimulus is found also with a mutated Av17 (mAv17, no proteinase inhibitor function). On the other hand, first therapeutic studies with mAv17 in the asthma model point in the direction that AHR is differentially regulated by cystatins with or without inhibitor function.

In recent studies protease-activated receptors (PARs) have been suggested to provide a novel pathway by which proteases affect innate immune responses (Shpacovitch 2007). The G-protein-coupled PARs can be activated by proteases via proteolytic cleavage and are found on epithelial cells, endothelial cells and leukocytes. Bacterial proteases and allergens with protease-activity have been shown to be able to activate distinct PARs and to be involved in allergic diseases (Reed 2004). Thereby, it might be possible that cystatin treatment addresses PARs by inhibiting their activation because proteases are blocked. In addition, eosinophils and mast cells were shown to express PARs on their cell surface. Addressing these receptors leads to release of mediators and thereby pathologic changes in tissues (Bolton 2003, Dugina 2003), which might also be prevented by cystatin treatment.

Filarial cystatin has been shown to efficiently block cathepsins involved in antigen processing (Schoenemeyer 2001). In this context it might be quite promising to investigate Av17 effects on dendritic cells (DC) in detail. First studies (diploma thesis of J. Hagen, 2007), revealed a phenotype characterized by $MHCII^{low}$ $CD40^{low}$ and $CD80/86^{int}$. Hence, they closely resemble immature DCs, although IL-12 secretion could be detected. Immature DCs or DCs that do not get proper maturation signals after antigen procession are described to induce anergy in T cells and thereby may lead to induction of tolerance (Lutz 2002). It is possible that cystatin achieves this immature DC state by interaction with cystatin C and by this, inhibition of the cysteine protease cathepsin S (needed for antigen procession and transport to cell surface). Still no information is so far available on the priming of T cells by Av17 pulsed DC. Furthermore, Av17 effects on other APCs, as macrophages, reveal a completely different picture with induction of TNF-α followed by massive IL-10 production (Hartmann 2002). Interestingly, in general, stimuli associated with Th2 response are often less strong inducers of DC maturation (DC2) than the ones associated with Th1 response induction (DC1). Immature DCs may induce tolerogenic immune response and induce T cells to become Tregs (Mahnke 2002). So far only a component of *Schistosoma mansoni* (lyso-PS) has been shown to effectively stimulate DCs to make IL-10 producing Tregs (van der Kleij 2002).

4.1.8. Mechanisms of immunomodulation by other helminth components

A. viteae cystatin seemingly provides a novel mechanism of immune modulation, which is different from other well-described helminth immunomodulators. In *Schistosoma mansoni* eggs a phosphatidylserine with specific chain length is expressed. It engages TLR2 expressed on dendritic cells, which in turn induces IL-10–producing regulatory T cells (van der Kleij 2002). Furthermore, schistosomes express 'self glycan' antigens that are recognized by lectin-receptors on host DCs, whose principal function is thought to capture self-glycan antigens and generate regulatory T cells to induce tolerance to these antigens. Thereby self-glycan antigens of schistosomes may deceive the host immune system to their own benefit. The host protects itself against damage by down-regulating helminth-induced Th2 immune responses, and may thus simultaneously be protected against excessive Th2 cell-mediated allergic responses (van Die 2006).

In addition *S. mansoni* eggs secrete a protein into host tissues that binds certain chemokines and inhibits their interaction with host chemokine receptors and their biological activity (smCKBP), thereby suppressing inflammation in several disease models (Smith 2005).

Only recently, a secreted glycoprotein of the filariae *A. viteae* (ES-62) was described to inhibit FcγRI-mediated mast cell responses by forming a complex with TLR4. This resulted in the sequestration of protein kinase C-α, a molecule important for mast cell activation via coupling of FcεRI to phospholipase D (Goodridge 2005b, Melendez 2007). Application of ES-62 could stop mast cell-dependent hypersensitivity in skin and lung. In addition, ES-62 has effects on B and T cells and APCs (DC, MØ) (Wilson 2003, Marshall 2005, Whelan 2000, Goodridge 2001). Furthermore ES-62 was shown to suppress collagen-induced arthritis (CIA) (McInnes 2003) and this was done at least partly by disruption of B cell activation and modulation of Toll-like receptor signaling on DCs and macrophages, influencing their maturation and cytokine production and not dependent on IL-10 (Goodridge 2004, 2005b).

The group of deMacedo, working on a protein extract of the tape worm *Ascaris suum* (PAS-1), was able to show that their component inhibits allergic airway inflammation (Lima 2002, Itami 2005) and anaphylaxis (deMacedo-Soares 2007) and that immunosuppression is dependent on IL-4 and IL-10 (Souza 2004). PAS-1 also led to IL-10 (and in some settings to TGF-ß) induction and could prime for regulatory T cells (deMacedo-Soares 2007). In addition PAS-1 treated macrophages produce IL-10 and TGF-ß (Itami 2005).

Further studies on the molecular mechanisms of immune modulation by filarial cystatin, such as the signaling pathways involved and the domains of cystatin responsible for the described effects are currently under way. Preliminary results indicate that mitogen-activated protein kinases are involved in the cystatin-induced modulation of macrophages, which is in line with recent data showing that inhibitors of mitogen-activated protein kinases may be suitable targets for anti-inflammatory therapy of asthma (Chialda 2005).

In conclusion, the data obtained on Av17 application in an asthma model, clearly suggest that the broad-acting anti-inflammatory activity of cystatin, which supposedly protects the worms from inflammation triggered by worm-derived components, might be exploited for the treatment of immune-related diseases. It selectively mimics the advantageous properties of a parasite infection without entailing its undesired side effects. Furthermore, data indicate that its protective effect is mediated by IL-10 secreting macrophages, which might be of the alternatively activated type II macrophage family (Figure 44).

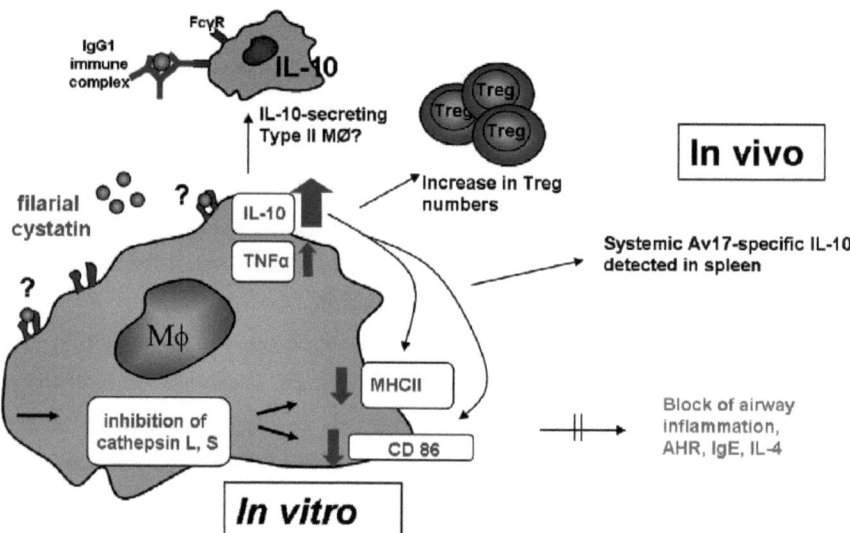

Figure 44: Effects of filarial cystatin (Av17) *in vivo* in a murine model of asthma linked to state of the art *in vitro* data: Recent studies revealed inhibition of cathepsins L and S, down-modulation of MHC class II (MHCII) and costimulatory molecule CD86 as well as strong induction of interleukin 10 (IL-10) and a short induction of tumor necrosis factor α (TNF α). The present *in vivo* data document a complete block of airway inflammation, airway hyperreactivity (AHR), allergen-specific IgE and local and systemic allergen-specific IL-4. Furthermore a systemic induction of cystatin-specific IL-10 was found (Av17-specific IL-10). In addition cystatin treatment led to increase in Treg numbers and it is hypothesized that macrophages are alternatively activated by TNF-α and Av17-IgG1-immune complexes binding to Fcγ receptor to become IL-10-secreting type II macrophages.

4.2. Influence of filarial cystatin on atopic dermatitis

Filarial cystatin could be demonstrated to provide protection of allergic airway inflammation in a murine asthma model, probably via IL-10 induction in macrophages (Schnoeller 2008). Application of cystatin in a murine model of atopic dermatitis (A.D.) now was moving from a strictly Th2 dependent allergy model to a mixed Th1/Th2 model. The concept of the pathobiology of atopic dermatitis involves a systemic Th2 response in addition to a biphasic T cell response in the skin with Th2 cells in acute A.D. and Th1 cells in the chronic phase

(Leung 2004). Application of filarial cystatin mediated profound amelioration of eczema formation in treated mice.

First, the clinical skin score, indicating the severity of inflammation, was nearly abolished after cystatin treatment. The stop of inflammation was also reflected by decreased epidermis thickness and, more importantly, by significantly decreased numbers of CD4 T cells and mast cells in skin. CD4 T cells are strongly linked to atopic dermatitis, mediating the inflammation by secretion of the cytokines IL-4 mainly in the acute phase but also IFN-γ and TNF-α in the chronic phase of the disease (Fiset 2006). Thereby downmodulation of these cells reflects an improved inflammation on cellular level. Mast cells are known to express FcεRI, bind IgE and degranulate in response to allergen, thereby releasing histamine, cytokines and chemokines. Mast cell degranulation in skin leads to erythema, edema and itch (Prussin 2006), which would also promote skin inflammation (Matsuda 1997, Dahten 2008). Therefore, the reduced mast cell numbers found after cystatin-treatment clearly point to a block of cutaneous inflammation. Interestingly mast cells are described to produce IL-25 (Ikeda 2003), a member of the IL-17 family, which is supposed to be strongly involved in atopic dermatitis (Toda 2003, van Beelen 2007). The role of IL-17 in Av17-treatment will be investigated in detail in the future.

Apart from differences in cell recruitment, sensitization was clearly affected by application of cystatin, leading to decreased levels of allergen-specific IgE. Although this did not result in decreased degranulation of RBL cells, in contrast to what was found in the asthma model, lower IgE levels and lowered mast cells numbers could still account for less allergic inflammation. However, this might also indicate that in this model of atopic dermatitis blocking IgE is not sufficient to suppress eczema, especially as a decrease in OVA-IgE without mast cell down-regulation, as seen with control protein DHFR, led to aggravation instead of amelioration. Nevertheless, in the model used in this work, the sensitization phase was described to be essential for induction of dermatitis (Dahten 2008) emphasizing the role of IgE in the early phase of inflammation.

Furthermore, cystatin application clearly altered cytokine patterns in systemic and local analyses. Allergen-specific cytokine production was systemically down-modulated, accounting for a lowered Th2 response, and no Th1 cytokines were measurable, arguing against a mere switch to Th1 to stop Th2 in the early phase (Wang 2007). Local cytokine measurements in patched skin pointed out, that the dermatitis-associated cytokines IL-4, IFN-γ and IL-10 were strongly downmodulated after cystatin treatment but anti-inflammatory TGF-ß was nearly 20 fold upregulated. These findings were in contrast to Av17-treatment in the asthma model, as TGF-ß was downmodulated in BALF and described to be detrimental in lungs, pointing to different effects of the anti-inflammatory cytokine depending on the model and the site of inflammation, respectively. However, Lan and co-worker could show that TGF-β in A.D. is produced by keratinocytes after tacrolimus treatment and thereby presumably involved in down-regulation of inflammation in eczema regions (Lan 2004). Apart from that, TGF-ß was found to be responsible for curing of skin lesions in the murine NC/Nga atopic dermatitis-model (Sumiyoshi 2002). Hence, such strong increase of TGF-ß in skin of Av17-treated mice seems feasible to mediate the protective effect but this will have to be proven in the future by application of TGF-ß-blocking antibodies.

Another aspect of cystatin treatment in atopic dermatitis might be Tregs. However, Treg numbers in skin-draining inguinal lymph nodes (ILN) were not altered by cystatin application. Interestingly, the Treg numbers in mesenteric lymph nodes were found to be "restored" to naïve levels. The role of Tregs in atopic dermatitis is contradictorily discussed. There are studies indicating a clear lack of $CD4^+CD25^+Foxp3^+$ Treg cells in human A.D. skin samples (Verhagen 2006), whereas other studies claim that Treg numbers are even increased in human eczema but loose their immunosuppressive capacity (Ou 2004). However, in the treatment with cystatin, Tregs do not seem to be the main mediators of protection, as Tregs in mesenteric lymph nodes will probably not directly improve skin inflammation and Foxp3 mRNA measured in eczema regions gave no clear hints towards an increase or decrease. Nevertheless, cystatin seems to be capable to restore at least partly, in several compartments, a possibly disrupted Treg network.

Apart from increase of TGF-ß in skin tissues other mechanisms of cystatin seem feasible. In the asthma model our group could show that cystatin effects are strongly dependent on IL-10 producing macrophages (Schnoeller 2008). Unfortunately depletion of macrophages over 10 weeks (duration of the dermatitis model) can hardly be achieved without the danger of animals dying due to infections, and more importantly altering the model. Antibodies against IL-10 receptor are also not feasible in this model, as local IL-10 reduction (in skin) improves inflammation itself (Sakamoto 2004) and a 10 week treatment again would be too "unhealthy" for the animals. Nevertheless, it is still possible that cystatin-specific IL-10 (which was found in spleens) downmodulated the *systemic* inflammatory response in a similar way as in the asthma model, without increasing IL-10 mRNA in skin, as IL-10 was also not increased in BALF.

Another feature of cystatin that has to be taken into account regarding atopic dermatitis is its proteinase inhibitor capacity. Many allergens are proteinases (for example the main allergen of house dust mite Derp1 is a group 1 allergen, a cysteine proteinase, Thomas 2002). Already starting with sensitization it has been shown that enzyme activity can be very important for the efficiency of this process (Harris 2004). A protease inhibitor applied in a Derp1 sensitization model led to inhibition of CD 23 cleavage from human B cells and significantly reduced HDM-induced permeabilization of the epithelial barrier (John 2000). Other cystatins, like cystatin A derived from human sweat, were shown to inhibit Derp1 induced IL-8 expression by human keratinocytes, thereby stopping its allergenic potential (Kato 2005). However, in this model OVA was used as allergen and ovalbumin is not a proteinase but a non-inhibitory serine protease inhibitor. Nevertheless, as the dermatitis model does not function under sterile conditions (personal communication A. Dahten), other allergenic components might be important, and they could indeed include proteases.

Recently the cytokine IL-31 was associated with atopic dermatitis. Bilsborough and colleagues could show that IL-31 is not only expressed by CLA (cutaneous lymphocyte antigen) positive T cells but that expression in keratinocytes and macrophages also correlates with severity of disease (Bilsborough 2006). It is quite possible that cystatin blocks IL-31 secretion by altering / differentially activating macrophages or blocking secretion in CLA-T cells but these analyses have to be done in the future.

Interestingly, another A.D. model makes use of IL-4 transgenic mice, in which the cytokine is expressed in the basal epidermis, using a basal keratinocyte-specific keratin 14 promoter (Chen 2004). It is possible that the downregulation of allergen-specific IL-4 found in Av17 treated animals (as also observed in the asthma model) contributed to amelioration. This might be due to the fact that IL-4 can prevent apoptosis of inflammatory cells in the dermis, as it was shown by Trautmann et al (2001), and thereby inflammatory cells persist.

Another interesting parameter associated with atopic dermatitis, and as well as with asthma, is TSLP (thymic stromal lymphopoietin). This IL-7-like cytokine is expressed by keratinocytes, triggers DC-mediated Th2 responses and could be correlated with severity of A.D. (Liu 2006). However, so far cystatin treatment did not lead to differences in TSLP-staining of skin sections.Taken together this work provides evidence that cystatin application in a murine model of dermatitis clearly abolishes eczema formation in challenged skin. The most outstanding alterations were found on cell recruitment and cytokine production. Av17-treatment induces strong up-regulation of transforming growth factor-ß (TGF-ß) in patched skin revealing a probable mechanism of amelioration (Figure 45). However, so far no defined cell population could be linked to the enhanced TGF-ß production. Thus it is quite possible that, as in Av17-treatment in asthma, Av17-specific IL-10 production and thereby differential

Discussion

cell recruitment and reduced allergen-specific Th2 cytokine secretion could have led to cure eczema, leaving TGF-β increase as a result of induced wound healing.

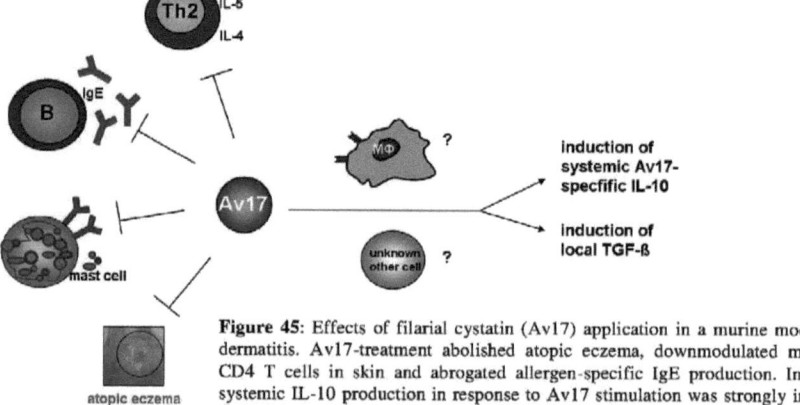

Figure 45: Effects of filarial cystatin (Av17) application in a murine model of atopic dermatitis. Av17-treatment abolished atopic eczema, downmodulated mast cell and CD4 T cells in skin and abrogated allergen-specific IgE production. In cured mice systemic IL-10 production in response to Av17 stimulation was strongly increased and TGF-β levels in skin dramatically upregulated. However, so far it is not clear if these effects are mediated by IL-10 producing macrophages or by unknown cells secreting IL-10 or TGF-β.

4.3. Worm infection and allergy: Comparing the influence of *H.polygyrus* in an asthma and a dermatitis model

Infections with parasitic worms have been demonstrated to potentially suppress allergic and inflammatory immune responses (Maizels 2004, Fallon 2007). However, there have also been some reports on aggravation of allergy in human patients and laboratory animals, which were so far linked to chronicity and severity of infection. Mouse models indicated, that only acute infection with *Trichinella spiralis* protects in a murine model of DTH, whereas a chronic infection with *H. polygyrus* does not provide amelioration. Interestingly in both infections strong IL-10 production was found, pointing in the direction that effects depend on the context of its release (Boitelle 2005).

Still, different outcome of parasite infection and allergy might not only depend on state (chronic or acute) and worm species but also on allergic disease and timing of helminth exposure (in utero, in first year, in adults) (Cooper 2006) as well as biology or localization of the parasite.

The present work provides evidence that modulation of immune responses by the gastrointestinal (GI) nematode *H. polygyrus* seems to be restricted to mucosa-associated tissues. Thus, infecting animals in an asthma (mucosa) and a dermatitis (cutaneous) model revealed a beneficial effect of the GI nematode *H. polygyrus* on airway inflammation but not in allergic skin lesions. *H. polygyrus* is a natural murine parasite with strictly enteric life cycle. 24-72h after inoculation third-stage larvae enter the wall of the small intestine and migrate into the muscularis externa beneath the mucosa. After 8-9 days adults emerge into the lumen and establish chronic infection in many strains of mice (Gause 2003). *H. polygyrus* induces a Th2 polarized immune response particularly in the MLN, blood eosinophilia and elevated levels of polyclonal serum IgE and IgG.

In the present work, infecting mice with *H. polygyrus*, interfered with allergic airway disease probably using several mechanisms: *H. polygyrus* infection significantly reduced the recruitment of inflammatory eosinophils into the lungs, thereby the release of effector molecules that lead to tissue damage (Trivedi 2007, Rothenberg 2006) can be prevented. However, total cell numbers in BALF were not significantly altered, which was due to an increase in macrophages in the worm–infected animals. The increase in macrophages might have important functions in the amelioration of inflammation by helminths, as it is shown that macrophages provided protection in a colitis model (Smith 2007). Furthermore also Mangan and colleagues observed an increase in BAL macrophages within worm infection in an ameliorated asthma model (Mangan 2006). However, depletion of macrophages was not performed in our model, as macrophage depletion in Mangan´s setting did not lead to reversion of protection by *Schistosoma mansoni* males (Mangan 2006).

In contrast to the asthma model, *H. polygyrus* infection in the dermatitis model did not interfere with induction of eczema, reflected by high clinical skin scores and increases in thickness of epidermis. Still in skin sections, decreased cell numbers in skin lesions of OVA-treated and *H. polygyrus*-infected (OVA/Hp) mice, with respect to $CD4^+$ and $CD8^+$ T cells, were detected. In contrast, mast cells showed significantly rising infiltrations into the eczema region in OVA/Hp-animals. The recruited mast cells probably contributed to the aggravation of eczema in OVA/Hp-mice, as they release soluble mediators involved in allergic reactions (Averbeck 2007). In this regard, it is interesting to note that mast cells are described to produce IL-25 (Ikeda 2003), a member of the IL-17 family, which is supposed to be strongly involved in atopic dermatitis (Toda 2003, van Beelen 2007). The interleukin 17 family consists of at least 5 homologues to the primarily described IL-17A (Cooke 2006, Zaccone 2008). IL-17A (or simply IL-17) is considered to be a proinflammatory cytokine. It is secreted by CD4 T cells, NK, CD8 and γδ-T cells. IL-17A and IL-17F (IL-25) are specifically linked to the new Th17 lineage, which has a role in MS (multiple sclerosis), RA (rheumatoid arthritis), SLE (systemic lupus erythematosus) and psoriasis. IL-25 is produced by activated Th2 cells (Fort 2001) and mast cells (Ikeda 2003). It can induce the expression of chemokines

Discussion

important for eosinophil and basophil recruitment (Fort 2001, Hurst 2002, Pan 2001), thereby amplifying the Th2 responses. Inhibiting IL-25 by soluble IL-25 receptor inhibits antigen-induced eosinophil and $CD4^+$ T cell recruitment into airways (Tamachi 2006). Administration of IL-17 members A, B, C, D and F leads to neutrophilia. IL-17D (IL-27) was first classified as a member of the IL-12 family. Its major function seems to be to limit Th17 effectors (Batten 2006, Weaver 2007).

Koga and colleagues could observe that human A.D. patients had increased Th17 cells numbers secreting IL-17 in peripherial blood and associated with severity of disease. In addition, IL-17 producing cells were found to be enriched in eczema skin sections (Koga 2008). In future studies IL-17 secretion and Th17 cells will be important and exciting features to investigate, especially as worm infections *(trichuris)* and gut immunity have already been linked to IL-17E (= IL-25) (Owyang 2006).

Hence, recruitment of cells differed significantly in the two allergy models and could reasonably have accounted for the different outcomes of disease progression.

Furthermore, *H. polygyrus* infection showed significantly decreased levels of allergen-specific IgE in both allergy models, probably resulting in less efficient sensitization of mast cells and basophils (Bradding 2006). However, the degranulation of a basophil cell line (RBL) sensitized with sera of OVA/Hp-mice was not reduced. This might be due to extremely increased total IgE levels in the OVA/Hp-group in the asthma and dermatitis model. Total IgE might have led to unspecific cross-linking of IgE receptors on RBL cells. Another explanation might be that OVA-specific IgE in the dermatitis model was bound by mast cells that massively infiltrated the skin and was therefore not measurable in sera. Thus, the decreased numbers of OVA-IgE in the dermatitis model might be due to a higher proportion of cell-bound IgE and not lowered production of IgE by plasma cells.

Interestingly, others did not find reduced OVA-specific IgE levels by nematode infections but still clearly ameliorated airway inflammation (Wilson 2005, Kitagaki 2006). Wilson and Kitagaki investigated settings with either completely established infection, meaning adult worms at the time point of first sensitization, or infection after sensitization phase. In contrast, in the present work mice were infected on the day of first sensitization. Therefore, it can be presumed that the inhibition of allergen-specific IgE is not essential for the amelioration of allergy by parasitic nematodes and that it is dependent of the time point of infection if allergen-specific antibody levels are decreased.

Another important point is the influence of *H. polygyrus* infection on cytokine production. Comparing both models, the Th2 cytokines IL-4 and IL-10 in response to allergen-stimulation were systemically decreased in spleens of OVA/Hp-mice in the asthma model. In contrast, spleen cell cultures of the OVA/Hp-group in A.D. did not show downregulation of allergen-specific IL-4, IL-5 or IL-10. Interestingly, both models showed increased systemic IL-4, IL-5 and IL-10 levels in response to parasite-antigen stimulation. Hence, only in the asthma model allergen-specific Th2 cytokine production was inhibited and in both approaches, *H. polygyrus*-infection resulted in an overall parasite-specific Th2 milieu. It is possible that the downregulation of allergen-specific IL-4 found in the asthma model (and also observed in Av17 treated animals in the asthma model) contributed to amelioration. In contrast, in the A.D. model, upregulated IL-4-levels could prevent apoptosis of inflammatory cells in the dermis and thereby inflammatory cells might persist (Trautmann 2001), accounting for aggravation of inflammation.

One key mediator responsible for the downregulation of asthma might be the anti-inflammatory cytokine IL-10. Gastrointestinal helminths are able to induce mucosal T cells (CD4, CD8 and also lamina propria cells) to make Th2 cytokines but also IL-10 and TGF-ß (Weinstock 2006). It has been shown, that systemic production of IL-10 can interfere with the development of airway hyperresponsiveness and allergic inflammation (Romagnani 2004, Fu 2006). Thereby, IL-10 modulates Th2 responses by suppression of allergen-specific IgE production and concomitant induction of non-inflammatory antibody isotypes, by reduction of pro-inflammatory cytokines released by mast cells, basophils and eosinophils and by indirect interference with Th2-associated phenomena such as mucus production (Urry 2006, Taylor A 2006). Thus, IL-10 is capable of redirecting pathologic allergic responses by a broad range of suppressive mechanisms. The importance of the constantly secreted IL-10 in helminth-driven modulation of allergic responses has already been shown in reports of infection with *H .polygyrus* or the blood fluke *Schistosoma mansoni* (Kitagaki 2006, Mangan 2004). However, systemic parasite-specific IL-10 production in the dermatitis model could not inhibit inflammation in skin. It is possible, that in this model the, relative to the asthma model, lower Hp-specific IL-10 levels were not sufficient to ameliorate disease but did alter infiltration of cells as indicated by reduced CD4 and CD8 T cells in skin.

However, it is also cogitable that the IL-10 producing cells could not migrate to the "right" site (draining lymph node) or, as indicated in the Av17 dermatitis experiments, IL-10 is not as important as TGF-ß levels in patched skin. In addition, it has been described that IL-10 in skin lesions may be detrimental instead of ameliorative (Simon 2007, Sakamoto 2004).

As mentioned before, TGF-ß has also been described as a potent mediator of immunomodulation by helminths (Maizels 2004, Dittrich 2008). Therefore, active TGF-β was measured at the site of inflammation either in BALF in the asthma model or TGF-β1 mRNA in skin tissues in the dermatitis model. In both approaches TGF-β levels were decreased in the helminth infected-groups. With regard to the asthma model, the data could indicate reduced fibrosis or already completed wound healing in lungs of OVA/Hp-mice rather than a lack of mediator, as it was also observed with Av17 treatment. This notion is supported by data on asthmatic individuals which clearly show increased TGF-β levels in the lung (Vignola 1997, Kenyon 2003). In contrast, in atopic dermatitis TGF-β is produced by keratinocytes after tacrolimus treatment and thereby presumably involved in downregulation of inflammation in eczema regions (Lan 2004). Furthermore, TGF-ß was described to be the factor that cured skin lesions in the murine NC/Nga atopic dermatitis-model (Sumiyoshi 2002). Hence, these data lead to the conclusion that reduced TGF-β levels might favour increased eczema in OVA/Hp-mice whereas they might indicate less fibrosis in the asthma model.

Several studies provide evidence for a role of regulatory T cells in helminth-induced amelioration of allergic airway hyperreactivity (Wilson 2005, Kitagaki 2006, Leech 2007). In accordance with a previous study on the role of Tregs in *H. polygyrus* infection (Rausch 2008) the present study points out, that the frequency of effector/memory-like Treg cells increased significantly in the mesenteric lymph nodes draining the site of helminth infection in both models of allergy. This regulatory T cell subset is characterized by the expression of the integrin $\alpha_E\beta_7$ (CD103) and was shown to exhibit a particular high suppressive capacity (Lehmann 2002), especially in the context of *H. polygyrus* infection (Rausch 2008). Such activated Treg cells efficiently control inflammation in models of colitis, delayed type hypersensitivity and arthritis and migrate preferentially to inflamed tissue (Huehn 2004).

Interestingly, in the asthma model a significant increase of such $CD103^+$ Tregs in the peribronchial lymph nodes draining the lung was detected, but no increase of Treg numbers in the skin draining inguinal lymph nodes in the dermatitis model was observed. Of note, it has been described recently that the induction of organ-selective Treg homing strongly depends on the site of priming (Siewert 2007). Infections alter the manner in which allergen-reactive cells migrate and localize to sites such as bronchial mucosa, a process highly dependent on the suite of chemokines, leukotrienes and prostaglandins (Luster 2004). Gut parasites might interfere with host T cell localisation or release specific agonists of these mediators and/or their receptors (Maizels 2005). Hence, one might speculate whether effector/memory-like Tregs induced or expanded by a concomitant nematode infection within the mesenteric lymph nodes also migrate to other inflamed mucosal tissues (and the relevant draining lymph nodes) and thereby interfere with allergic disorders, as suggested by the data from the asthma model.

In contrast, such parasite-related Treg cells might fail to enter non-mucosal compartments, such as the skin, due to their priming in mucosa-associated lymphoid organs and the lack of homing receptors needed to enter such sites. The obtained data on lowered frequencies of effector/memory-like Tregs in skin-draining lymph nodes and absence of Foxp3-expression in the inflamed skin in mice infected with *H. polygyrus* strongly support this view, however further investigation is needed. Still, Hirahara and co-workers described, that in healthy individuals nearly all Foxp3$^+$ Tregs within PBMCs express skin homing receptors like CLA (cutaneous lymphocyte antigen, Hirahara 2006) and Iellem could proof that usually natural Tregs express much more skin-homing than gut-homing receptors (Iellem, 2003) implying that Tregs in healthy individuals more likely migrate to the skin than to mucosal tissue. Worm infection might trigger large numbers of Treg cells to mucosal tissue or alter intergrin expression patterns (Weinstock 2006). In this account, very recently it was shown by Ozdemir and colleagues that transfer of CD4$^+$ MLN of mice with intestinal immediate type hypersensitivity could induce airway hyperreactivity (Ozdemir 2007). They conclude that cells are able to migrate to other mucosal tissue. This would go in line with modified MLN from *H. polygyrus* infected animals being able to mediate suppression of mucosal airway but not skin inflammation.

Nevertheless one has to keep in mind that the role of Tregs in atopic dermatitis is not clear: Verhagen (2006) observed a clear lack of CD4$^+$CD25$^+$Foxp3$^+$ Treg cells in human A.D. skin samples, whereas Ou and colleagues (Ou 2004) claim that Treg numbers are even increased in human eczema but loose their immunosuppressive capacity.

Taken together, the data obtained in this work suggest a critical role for Tregs at the site of infection. Tregs induced or activated in the context of the nematode infection might express a homing receptor pattern beneficial for the entry to the inflamed mucosal site and thereby impair further development of airway inflammation. These alterations might also account for unchanged or even exacerbated dermatitis, as Tregs might not be able to enter cutaneous compartments and draining lymph nodes due to impaired homing abilities.

Infection with helminths (*Schistosoma mansoni*, hookworm; Araujo 2006, Leonardi-Bee 2006) has been reported to be negatively associated with human asthma. In contrast a human study analysing *Trichuris trichiura* infection indicated rather an aggravated than a reduced risk of A.D. in such patients (Haileamlak 2005). In a recent study in Cuba it turned out that *A. lumbricoides* infection (life cycle with migration through host) protected against atopic dermatitis in children whereas infection with *Enterobius vermicularis* (strictly enteric life cycle) was associated with increased risk of atopic dermatitis (Wördemann 2008). These

Discussion

studies strongly suggest that the species and localization of the parasite might be crucial for ameliorative or detrimental effects on different allergic diseases. In this context, future studies with skin-dwelling nematodes (as *Acanthocheilonema viteae* or *Onchocerca volvolus*) might reveal new mechanisms regarding allergic skin-disease and also help to unravel the positive effects on A.D. found with *A. viteae* cystatin.

In summary, this study with the mouse nematode *Heligmosomoides polygyrus* provides explanations for the discrepancy among asthma and atopic dermatitis. A lack of Tregs at the site of inflammation in the atopic dermatitis model and a clear boost of Treg numbers in the draining lymph nodes in the asthma model are indicative. Hence, elementary differences in immune cell recruitment and cytokine production due to the nematode infection in different disease settings were observed and therefore intestinal helminth infections seem not to be thoroughly useful to improve allergic reactions (Figure 46). In this respect defined parasite components, as the detailed described cystatin, will be a better option to take advantage of the modulatory capabilities of helminth infections.

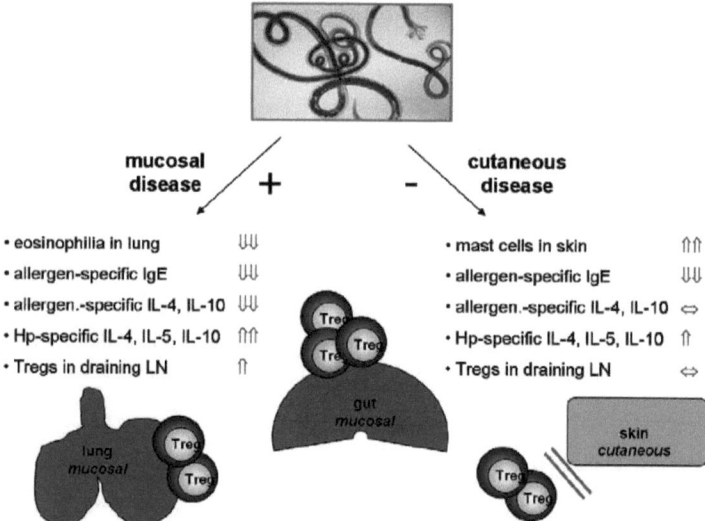

Figure 46: Differential outcome of *H. polygyrus* infection in two models of allergic disease. Mucosa-associated airway inflammation is ameliorated by *H. polygyrus* infection, indicated by strongly decreased eosinophilia in lung, ablated allergen-specific IgE antibodies and Th2 cytokines. Parasite-specific Th2 cytokines and more importantly regulatory T cells are clearly increased, resulting in a modified Th2 response and downregulated disease. In contrast, cutaneous disease is not improved by infection with the gastrointestinal parasite. Mast cells in challenged skin are increased, allergen-specific Th2 cytokines are not downmodulated and no increase in Treg numbers in the draining ILN is found. This results in inflammation. Furthermore Treg cells induced in the gut seem not to be able to migrate to the skin but to migrate into lung mucosa.

5. Outlook

Filarial cystatin has been shown to bear great potential to interfere with distinct allergic diseases and macrophage-mediated colitis (Schnoeller 2008). In the future it will be of great interest to apply cystatin in more clinically relevant allergy models, such as the birch pollen allergen Betv1 or house dust mite allergen Derp1, in order to prove cystatin's applicability in the clinics. Furthermore, other disease model as DTH (delayed type hypersensitivity, EAE (experimental autoimmune encephalomyelitis) or SLE (systemic lupus erythematosus) should be tested to determine the priority field of application.

One important aspect will be to make sure that cystatin treatment does not impair immune responses against other pathogens. This might be done in listeria or influenza infections "challenged" with cystatin applications.

Another aspect to be tested is the exact mechanism of cystatin action in vivo. Receptor screening and gene expression analyses to identify the cellular receptor and the signalling cascade are planned.

In addition, B cells and dendritic cells should be monitored to gain more information on interaction partners and cellular networks employed. This might be done by transfer experiments, in knock-out mice or by specific cell depletion.

Finally, studies to exclude cross-reactions of filarial cystatin with human allergens have to be carried out by IgE ELISA of sera of atopic patients.

Concerning worm infection and allergy, future studies have to concentrate on parasite biology and localization. Helminths associated with cutaneous tissue should be investigated in distinct disease models. Furthermore, experiments to analyze the migration pattern of gut-induced versus skin-induced Tregs will provide further insight on the underlying mechanisms.

6. Material and Methods

6.1. Molecular biology and protein biochemistry methods

6.1.1. Expression and purification of recombinant *Acanthocheilonema viteae* cystatin

A pET Av17 clone was incubated overnight at 37°C in 30 ml LB medium containing 50µg/ml kanamycin. 10 ml of overnight culture was used to start large scale cultures (300-500 ml), in which expression of the protein was induced by addition of 1mM IPTG isopropylgalactoside for 3 h when cultures reached an optical density OD_{600} of 0.6. After centrifugation (4 °C, 6,000 rpm, 15 min) pellets were resuspended in 10-20 ml lysis buffer containing lysozym and incubated 30 min on ice. Lysates were sonificated (3 min, 20 %) and centrifuged again (4 °C, 12,000 rpm, 15 min) to get protein containing supernatants. The protein was purified by affinity chromatography using Ni-NTA resin (Diagen, Hilden, Germany) and glycerol-PBS-buffer (10 % glycerol/PBS buffer; buffers B). Recombinant Av17 was eluted via pH gradient (pH6, 5, 3) subsequently neutralized by addition of neutralization buffer and dialyzed against PBS/0.05 % Triton X-100.

Each lot of rAv17 had to be tested for its biological activity. Therefore, the protein was analyzed for its suppressive effect on polyclonal proliferation of mouse splenocytes in response to concanavalin A. Furthermore, cysteine proteinase inhibitor function was tested in activity test to papain cleavage. Cystatin was incubated with papain and the substrate BAPNA and inhibition of substrate cleavage by papain was measured.

6.1.2. Purification of recombinant Cysele2, DHFR and mAv17

Cysele2 (*C. elegans* cystatin 2) was induced as described before (6.1.1.) and supernatants were purified using FPLC and urea buffer. Elution was performed with pH shift (pH 6.4, pH 5.9, pH 4.5). To express recombinant mouse dihydrofolat reductase (DHFR) the procedure was similar to Av17 except that bacterial cultures contained 50 µg/ml kanamycin and 100 µg/ml ampicillin. Furthermore recombinant DHFR was purified by an imidazol gradient (buffers A) using FPLC (fast performance liquid chromatography) and HiTrap FF columns (similar to Ni-NTA).

A mutated version of Av17 (mAv17) with point mutations in the active centres for cysteine proteinase inhibitor activity was purified similar to Av17 and DHFR, via NiNTA-matrix and pH shift.

Endotoxin concentration Av17: 2.5 pg / µg protein = 0.05 EU per application; DHFR: 2.8 pg / µg protein = 0.056 EU per application; Cysele2: 2.0 pg / µg protein = 0.04 EU per application; mAv17: 2.5pg / µg protein = 0.05 EU per application.

6.1.3. Endotrap system to remove endotoxin contaminations

To remove endotoxin contaminations, Endotrap red columns were used according to the manufacturer's instructions (Profos, Regensburg, Germany).

6.1.4. Limulus amoebocyte test (endotoxin measurement)

Final endotoxin concentrations were detected by limulus amoebocyte lysate (LAL) test (Cambrex, Bio Sciences, Walkersville, USA) according to manufacturer's instructions.

6.1.5. Quantification of protein

Protein concentrations were measured with the bicinchoninic acid test using BCA protein assay kit (Pierce) according to manufacturer's instructions (Pierce, USA).

6.1.6. SDS-PAGE (sodium-dodecyl-sulfate polyacrylamid-gelelectrophoresis) and coomassie staining

SDS-PAGE was used to determine expression, purity and size of induced proteins. SDS-PAGE separates proteins according to their size. Differences in polarization are mainly eradicated by denaturation with β-mercaptoethanol and heat and binding of anionic SDS to the proteins in constant ratios. According to the expected size of the wanted protein separation gels of 12-14 % polyacrylamid were used, always with a 6 % stacking gel. Gels were run at 120V/400mA. To visualize proteins gels were stained 30 min at 60 °C with coomassie staining solution. Unbound coomassie was washed away with de-staining solution.

6.1.7. Dialysis

Protein preparations were dialysed against PBS/0.005 % Triton over night and additional 2 hours next day in fresh buffer if not indicated differently.

6.1.8. RNA and cDNA preparation of lung and skin tissue and cultured cells

For RNA preparation tissues or cells were frozen in liquid nitrogen and then ground with ice-cold mortar and pistil. The powdered material was then treated according to manufacturer's instructions of RNeasy Mini Kit (Qiagen, Hilden, Germany) followed by digestion of DNA using the RNase-free DNase-Set (Qiagen) according to the manufacturer's instructions.

Material and Methods

RNA was reverse-transcribed using the TaqMan Reverse Transcription Reagent (Applied Biosystems, Warrington, UK) and oligo dTs. The PCR conditions were 10 min 25 °C, 30 min 48 °C, 5 min 95 °C.

6.1.9. Real time PCR (TaqMan-System)

The basic principle of real time PCR is the fluorescence resonance transfer (FRET). FRET is the phenomenon when two fluorochromes with overlapping adsorption and emission energy act as reporter and quencher. This means that if the two fluorochromes are in close contact the emission energy of the reporter will be transferred to the quencher and therefore the emission energy of the quencher and not of the reporter is set free. Real time PCR uses this phenomenon by the application of a gene-specific probe (oligonucleotide with reporter and quencher fluorochrome in close contact) that is destroyed by the 5´-3´-exonuclease activity of the polymerase when it amplifes the sequence. The more amplificates are produced the more reporter is set free and the fluorenscence signal increases. The amount of original mRNA can be recalculated and set into relation to a housekeeping gene as described below. Real time PCR was used to analyse the expression levels of several cytokines at the site of inflammation / in the tissue (mRNA levels). The 7300 Real-Time PCR System (Applied Biosystems, New Jersey, USA) and TaqMan reagents were used to do so. PCR amplifications were done in triplicates containing 3 µl of cDNA, 2 µl of 20x TaqMan labelled primer mix and 10 µl of 2x TaqMan PCR buffer. The 20x TaqMan primer mix consisted of two unlabeled PCR primers (900 nM each final concentration) and 1FAM™ dye-labelled TaqMan®MGB probe (250 nM final concentration). All primers were obtained from Applied Biosystems (IL-4 Assay ID: Mm00445259_m1, IL-10 Assay ID: Mm00439616_m1, TGF-ß Assay ID: Mm 00441729_g1, Foxp3 Assay ID: Mm00475156_m1, IFN-γ Assay ID: Mm00801778_m1, GAPDH Assay ID: Mm99999915_g1). PCR conditions were set to 10 min denaturation at 95°C followed by 40 amplification cycles of 15 s at 95°C and 60 s at 60°C. The relative amounts of cytokine mRNA were normalized to the endogenous reference, the house keeping gene GAPDH. Calculation of expression profiles was done using the $2^{-\Delta\Delta Ct}$ –method. In this method a threshold is set according to the cycle number where the individual reactions start to get exponential. The program calculates Ct-values (Ct = cycle threshold) for each sample, which reflect the cycle number at which each sample reaches the given threshold. Ct values of the endogenous control were subtracted from the correlating samples to normalize samples. Mean values of sample triplets were then compared to controls (either negative or positive control) and calculated as described earlier (Livak 2001).

6.2. Cell culture techniques

6.2.1. RBL assay

RBL assay was performed to analyse antigen-specific IgE in sera. RBL (rat basophile leukemia) cells were maintained in Eagle's minimum essential medium with 10 % RPMI, 5 % fetal calf serum (FCS), penicillin (100 U/ml), streptomycin (100 µg/ml) and L-glutamine (2 mM). Cells were plated in 96-well flat bottom plates (10^5 cells/well). Passive sensitization of the cells was performed by 1 h incubation with 50 µl mouse serum diluted 1:50 in culture medium. Cells were washed with Tyrode's buffer and for cross-linking 5µg/ml OVA or 5µg/ml Av17 diluted in Tyrode's buffer for 1 h was added. Spontaneous and total release was measured by incubation of cells with Tyrode's buffer without antigen and addition of 1 % Triton X-100 respectively. β-hexosaminidase release was measured in 30 µl supernatant and the enzymatic activity was detected by addition of 50 µl of the substrate p-nitrophenyl N-acetyl ß-d-glucosaminide (1.3 mg/ml in 0.1 M citric buffer, pH 4.5). The reaction was stopped with 100 µl 0.2 M glycine solution ph 10.7 and absorbance was measured at 405 nm. Total release was set to 100 % and results are expressed as percentages.

6.2.2. Preparation of mesenteric lymph node cells (MLNC), splenocytes, inguinal (ILNC) and peribrochial lymph node cells (PBLNC)

MLN, PBLN, ILN and spleen were isolated aseptically from euthanized mice. Organs were dissociated by passing them through a steel mesh in PBS pH 7.4 containing 0.2% BSA to obtain single cell suspensions. Erythrocytes were removed by resuspension of washed cells in erythrocyte lysis buffer for 5 min on ice. After washing, cells were adjusted to desired concentrations in cRPMI for culture or PBS/BSA for surface stainings and flow cytometry analysis.

In the asthma model spleen mononuclear cells were isolated by density gradient centrifugation (Lympholyte-M, Cedarlane Laboratories, Hornby, Ontario, Canada).

Cells were cultured with a density of 5×10^5 cells/well in RPMI 1640 (10 % FCS, Pen/strep, L-Glu and fungi) for 96 hours in the presence of 50 µg/ml OVA, 10 µg/ml Av17, 10 µg/ml DHFR or 2,5 µg/ml concanavalin A (ConA). Cell culture supernatants were stored at -20 °C until performance of cytokine ELISA.

6.2.3. Peritoneal lavage and preparation of peritoneal exudate cells (PEC)

To obtain PEC the peritoneal cavity of sacrificed mice was washed 4 to 5 times with 2 ml ice-cold RPMI. Cells were centrifuged and lysis of erythrocytes was performed. Afterwards cells were seed in 96 well plates at a density of 2.5 – 3.0 x 10^5 cells/well in complete RPMI. For some experiments medium was changed after 2h and cells were left to adhere another 12 hours before stimulation with antigen. Cell culture supernatants were stored at -20 °C until performance of cytokine ELISA.

6.3. Animal models

6.3.1. Animals used for experiments

Female BALB/c mice (Ol1a-Hsd) were purchased from Harlan Winkelmann (Borchen, Germany) or the BfR, Berlin respectively. For asthma experiments mice were kept in ivc-cages, for other experiments animals were housed with filter caps. All mice were housed and handled following national guidelines and as approved by the animal ethics committee.

6.3.2. Murine asthma model and measurement of airway hyperreactivity (AHR)

Female BALB/c mice (Harlan-Winkelmann, Bachem, Germany) were sensitized twice (day 0 and day 14) intraperitoneally with 20 µg OVA (grade VI, Sigma-Aldrich, Steinheim, Germany) emulsified in 2 mg of aluminium hydroxide (Imject®Alum, Pierce, Rockford, USA). On days 28 and 29, mice were challenged intranasally with 50 µg OVA (25µl per nostril). Airway responsiveness was measured in unrestrained and conscious mice on day 31. After provocation by inhalation with rising dosages of the bronchostringent metacholin (MCh), AHR was subsequently determined in a full body plethysmograph. The pressure difference between the plethysmograph chamber containing the animal and a reference chamber was assessed. This difference during the respiratory cycle can be indicated as enhanced pause (PenH) using mathematical formulae and can be taken as an index for respiratory constriction amongst the experimental animals (Hamelmann 1997c). A baseline of breathing was recorded as reaction to inhalation of PBS. These baseline values were subtracted from values obtained in response to stimulation with 6.25 mg/ml, 12.5 mg/ml, 25 mg/ml and 50 mg/ml metacholine. Measurements were performed for each bronchoconstrictor concentration for 3 minutes and values obtained were evaluated as follows: Data was filtered to ensure that only measurements with a success rate (Sr) of more than 50 % were included. Furthermore only Penh values that correlated with a pressure (EIP) lower than 10 and Te values bigger than Ti values were used for calculation. The mean values of the Penh (pause enhanced) for every MCh concentration was calculated and index Penh values of individual animals (Penh value of MCh minus baseline) were analysed as mean values of groups.

One day after assessment of airway hyperreactivity mice were sacrificed and the following analyses were carried out:

- lavage of the lung to assess BALF and cell numbers and types pointing to inflammation
- removal of spleen to perform cell culture and proliferation assays
- removal of lymph nodes (peribronchial (PBLN), mesenteric (MLN)) to perform FACS analysis and / or cell cultures
- cryoconservation of parts of the lung to perform Real-Time-PCR
- preparation of lungs for histological analysis (formalin or cryo)
- assessment of blood to monitor antibody production

Within the model 20 µg recombinant *A. viteae* cystatin in a total volume of 200µl diluted in low-endotoxin PBS or the same amount of control protein DHFR was injected intraperitoneally four times in weekly intervals during the sensitization (d 1, 7, 14, 21; preventive model), or three times after sensitization with ovalbumin prior to airway allergen challenges (d 21, 23, 25; pre-challenge model). Naïve control animals were treated with PBS in aluminium hydroxide intraperitoneally and challenged with PBS intranasally.

6.3.2.1. Bronchoalveolar lavage (BAL)

Animals were sacrificed by cervical dislocation and lungs were lavaged twice using a tracheal drain tube with 0.8 ml of cold PBS/EDTA (proteinase inhibitor tablets, (complete^{TM} Mini)). The volume of the lavages was measured by weighing and after centrifugation (10 min, 2.000 rpm, 4 °C) supernatants of the first lavages were stored at -20 °C for cytokine analyses. The cell pellets of both lavages were resuspended in a total of 1ml PBS, cell numbers were counted and 100µl of the cell suspensions was used to prepare cytospin slides. Cytospin slides were stained with DiffQuick (Dade Behring AG) and bronchoalveolar lavage (BAL) cells were differentiated by morphological criteria (count of 200 cells under light microscopy), as previously described (Blümchen 2006).

6.3.2.2. Histological analysis of lung

Lungs were either filled with Tissue Tek OCT compound (Jung) solution after removal and frozen in liquid nitrogen or put in 3.5 % formalin to be embedded in paraffin later on. Slices of 5µm were cut and afterwards stained with he (hematoxilin/eotaxin) to monitor cells types and infiltration in lung tissue or stained with PAS/H to visualize mucus production in bronchioles (Edwan 2004). Slides were analysed regarding bronchioles and their surrounding by light microscopy with a 40- and 400- fold magnification.

6.3.3. Murine dermatitis model (atopic eczema)

Mice were sensitized i.p. on days 1, 14 and 21 with 10 µg ovalbumin (OVA, Sigma-Aldrich) adsorbed to 10 µg aluminium hydroxide (Imject®Alum) in 100 µl PBS or with phosphate buffered saline (PBS) and alum as control. Epicutaneous (e.c.) allergen application was performed as described previously (Spergel 1998). Each mouse had a total of three one-week exposures to the same skin site. Naïve animals were treated e.c. with PBS/Alum. At day 70 mice were sacrificed and skin biopsies, spleen, lymph nodes and blood samples were collected for further analysis.

To evaluate the severity of OVA-induced eczematous skin lesions, standardized clinical skin score (CSS) based on clinical features of human A.D. (Leung 1990) was used, which considers typical skin features of A.D., like erythema, edema/papules, oozing/crusts, dryness and extension (Matsuda 1997). For total CCS the marked lesions that were visible on the patched skin area were graded as 0 (none), 1 (mild), 2 (moderate) and 3 (severe) by two independent persons. The total score was taken as index of dermatitis severity (Dahten 2008).

Within the model 20 µg recombinant *A. viteae* cystatin in a total volume of 200 µl diluted in low-endotoxin PBS or the same amount of control protein DHFR was injected intraperitoneally six times on day 1, 7, 14, 21, 42 and 63. Naïve control animals were injected with PBS on cystatin application days.

6.3.3.1. Histological analysis of skin

Skin samples were embedded in Tissue Tek OCT compound. Frozen tissue blocks were cut into 5 µm sections and fixed in acetone. For histological analysis sections were stained with hematoxylin for 1 min. Thickness of epidermis and dermis was determined by Axiovisions measuring-tools on the Axioplan light microscope (Zeiss) at 100 x magnification. Ten measurements per mouse were obtained and expressed in micrometers.

6.3.3.2. Immunohistochemistry of skin sections

Sections of 4 µm were prepared as described above, blocked with 10 % normal goat serum (Dako) and incubated with rat anti-mouse CD4 (clone RM4-5) and anti-mouse CD8 (clone 53-6.7) followed by incubation with biotinylated goat anti-rat IgG. CD11c positive cells were detected by usage of purified hamster anti-mouse antibody (HL3) and biotinylated anti-hamster IgG cocktail (all BD Pharmingen, Heidelberg, Germany). To rule out unspecific binding of antibodies, negative controls without primary and/or secondary antibodies were performed, respectively. Signals were detected by Chem-Mate® alkaline phosphatase/red detection Kit (Dako) and by hematoxylin counter staining. To determine infiltrating mast cells, 5μm skin sections were stained for 1h with o-toluidine blue. Positive stained cells were counted in 10 high power fields (HPFs) at x 200 magnification and expressed as cells per HPF.

6.3.4. Depletion of Treg cells and blocking of IL-10 receptor in vivo

Treg cells were depleted by intraperitoneal application of 100 μg anti-CD25 antibodies (clone PC61, gift A. Scheffold, DRFZ, Berlin) 5 days prior to challenge/analyses. Treg depletion was confirmed by flow cytometry surveying expression of CD4, CD25 and CD103 coexpression. An isotype-matched antibody (Rat IgG) was used as control.

Anti-IL-10 receptor antibodies (clone 1B1) were applied three times intraperitoneally 500 μg each time along with filarial cystatin. An isotype-matched antibody (Rat IgG) was used as control.

6.3.5. Depletion of macrophages / Preparation of multilamellar vesicles (MLV) containing clodronate

Clodronate liposomes were prepared as described by Van Rooijen elsewhere (Van Rooijen 1994). In brief, an emulsion of 86 mg/ml phosphatidylcholin and 8 mg/ml cholesterol in 10 ml chloroform was vacuum evaporated and multilamellar vesicles emerged by gentle shaking with 10 ml clodronate suspension (0.6 M, 0.25 mg/ml) or PBS under nitrogen conditions. The vesicles were kept under nitrogen gas (2 h) and sonicated in a waterbath sonicator for 3 min. The suspension was kept under N_2 over night for swelling of the liposomes. The liposome solution was centrifuged 15 min at 10,000 g and remaining clodronate removed. The clodronate was sterile filtered and reused up to 5 times to avoid spoiling. MLVs were washed with sterile PBS twice (30 min, 25,000 g) and resuspended in the same buffer. Macrophages were depleted by intranasal and intraperitoneal application of clodronate liposomes two days prior to challenges (100 μl per application). Macrophage depletion was confirmed by flow cytometry (see above) or histological cell differentiation (see BALF), respectively.

6.4. Immunological methods

6.4.1. Enzyme-linked immunosorbent assay

6.4.1.1. Serum levels of IgE (total and OVA-IgE), OVA-IgG1, OVA-IgG2a

Total IgE levels were measured by sandwich ELISA using anti-IgE antibody as catcher, 3% BSA in PBS to block unspecific binding, dilutions of sera (1:10, 1:100, 1:1000, 1:10000 in blocking buffer) containing induced antibodies and anti-IgE antibody coupled to Biotin as detection antibody. HRP (horse radish peroxidise) labelled Streptavidin was applied in a 1:10000 dilution to detect Biotin-labelled antibodies and binding was visualized by application of the HRP-substrate TMB. After stopping the reaction with 1M H_2SO_4 plates where measured at 450/630 nm in a plate reader. IgE-standard solution was analysed in accordance to samples to calculate IgE amounts.

OVA-specific antibodies where measured in an analogical manner though after application of diluted samples 3 µg/ml biotinylated ovalbumin was used to detect exclusively OVA-specific IgE antibodies. To calculate amounts, purchasable IgG1-standards or IgE-and IgG2a standard sera were used, respectively. OVA-specific IgE and IgG2a are therefore calculated as lab units (LU) and not in µg (Beier 2004).

immunoglobulin	Capture ab	standard	Detection ab
Total IgE	αIgE 10µg/ml PC284 The binding site	IgE 250ng/ml 557079 Pharmingen	αIgE-Bio 2,5µg/ml 553414 Pharmingen
OVA-IgE	αIgE 2µg/ml 553413 Pharmingen	OVA-Standardserum #67 1:64	OVA-Bio 3µg/ml
OVA-IgG1	αIgG1 2µg/ml 553445 Pharmingen	OVA-spec. IgG1 340ng/ml A6075 Sigma	OVA-Bio 3µg/ml
OVA-IgG2a	αIgG2a 2µg/ml 553387 Pharmingen	OVA-Standardserum #67 1:64	OVA-Bio 3µg/ml

6.4.1.2. Biotinylation of ovalbumin

1ml of a 1mg/ml solution of OVA grade VI (pH 8.3 with 100 mM $NaHCO_3$) was incubated 1h with 13.4 M sulfo-NHS-LC-Biotin (20 µl) at RT, shaking, in the dark. Afterwards the solution was purified of unbound biotin by size exclusion chromatography (application of PD10-sephadex column).

6.4.1.3. Subclass - specific ELISA

200 ng/well of the referring protein was coated in carbonate buffer overnight and plates were blocked with 3 % BSA/TBS. Sera were applied in 1:50 dilution (in 1 % BSA/TBS) and antibody subclasses were detected by use of subclass-specific AP(alkaline phosphatase)-

labeled detection antibodies and AP-substrate BCIP (5-Brom-4-chlor-3-indoxylphosphate) and NBT (Nitroblue-Tetrazoliumchlorid). Reaction was abrogated with EDTA and measured at 405/630 nm in ELISA reader.

6.4.2. Cytokine detection

IL-4, IL-5, IL-10, IFN-γ and TNF-α measurements were performed using OptEIA ELISA kits according to manufacturer's instructions, but using 50 µl of samples and standards/well. TGF-β was analysed by R&D TGF-β ELISA kit also according to manufacturer's instructions. To measure TGF-ß in cell cultures cells were cultivated in RPMI/1 % FCS/1mg/ml BSA in order to reduce background levels. In several experiments cytokine bead arrays to analyse G-CSF, IFN-γ, IL-1β, IL-2, IL-4, IL-5, IL-10, IL-12p40, IL-13, KC, RANTES, Eotaxin and TNF-α simultaneously were performed (mouse Bio-13-Plex assay) according to manufacturer's instructions.

6.4.3. Flow cytometric analysis

Surface staining of lymphocytes using mAb was performed in PBS/0.2 % BSA on ice and in the dark for 15 min. Cell suspensions (1×10^6 total cells) were washed in PBS/BSA and stained with αCD4, αCD25 and αCD103 to detected Treg cells. Macrophages in peritoneal excudate cells (PEC) were stained for αCD3 and αCD19 to exclude T and B cells and detected by αF4/80- staining. Unspecific binding of the mAbs was blocked by the addition of αFcγRII/III (20 µg/ml). Intracellular detection of Foxp3 was performed according to the manufacturer's instructions. Non-specific intracellular binding was blocked with whole rat IgG. Cytometric analysis was performed using FACS Calibur or LSRII and FlowJo software.

6.4.4. Proliferation assay

Standard proliferation assays were performed by culturing 2.5×10^5 splenocytes in 96 well plates. Cells were stimulated with protein or mitogen for 24h. Subsequently 1 µCi 3H-Thymidin per well was added for another 20 h. The uptake of labelled thymidin was measured by scintigraphy and reflects proliferation rate in cpm (counts per minute).

6.5. Parasitological methods

6.5.1. Life cycle of *Heligmosomoides polygyrus*

Heligmosomoides polygyrus was maintained by serial passage in BALB/c mice. Mice were infected with approx. 200 L3 using a feeding tube. Infective larvae were obtained from fecal cultures. Feces were collected from infected animals, washed in distilled water and plated in Petri dishes on humid blotting paper for 7 days. On d7, L3 were washed from the plate and kept in distilled water at 4 °C after extensive washing. L3 were used for infection up to 8 weeks after fecal culture. Mice in experiments were infected with a defined dose of L3. The adult worm burden was determined by collecting adult worms from the small intestine on the day of dissection.

6.5.2. Fecal egg count

To monitor the status of *H.polygyrus* infection feces were analysed for egg output regularly. Feces of individual mice were collected, weighed and suspended in 1 ml aqua dest. by a mortar. Subsequently 5 ml saturated NaCl-solution was added and 650 µl of the solution put onto MacMaster chambers to count eggs via light microscopy. EPG (eggs per g feces) was calculated by the following formulae:

Number of eggs in sample x volume x 6.67 x (1/ feces weight in g).

6.5.3. Preparation of *H. polygyrus* adult worm antigen

Soluble worm antigen was prepared from adult worms kept in culture in RPMI medium containing 100 U/ml penicillin and 100 µg/ml streptomycin for 24 h. Worm material was homogenized and sonicated (1 min, 60W) on ice in PBS (pH 7.4). The homogenate was centrifuged (20 min, 20.000g, 4 °C) and the supernatant was passed through a 0.4 µm filter for sterilization. Antigen extracts were stored at –80 °C until application.

6.6. Statistical analysis

Each experiment was performed with 5–6 animals/group, and the experiments shown are representative of two to three independent experiments unless stated otherwise. Statistical analysis was performed with the two-tailed Mann-Whitney U test. Data are presented as means +/- SE. Values of $p < 0.05$ were considered to be statistically significant.

6.7. Material

6.7.1. Laboratory equipment

Flow cytometer: LSRII	BD Biosystems, San Jose, CA
ELISA-reader	Dynatech, Denkendorf, Germany BioTek, Vermont, USA
Scintillation-spectroscope Trilux 1450	Wallac, Turku, Finland
Ultrasound-disintegrator	Heinemann, Schw. Gmünd, Germany
Real-time PCR system 7300	Applied Biosystems, Foster City, CA
whole-body plethysmograph (AHR)	emka technologies, Paris, France
Cytocentrifuge (cytospin)	ThermoShandon, Frankfurt, Germany

6.7.2. Buffers and media

FACS staining buffer	PBS / 0.2 % BSA
FACS fixation buffer	PBS / 0.2 % BSA / 2 % paraformaldehyde
Erythrocyte lysis buffer pH 7.5	0.01 M $KHCO_3$ 0.155 M NH_4Cl 0.1 mM EDTA
Cell culture medium (cRPMI)	RPMI-1640 5-10 % FCS 20 mM L-glutamine 100 U/ml penicillin 100 µg/ml streptomycin all from Biochrom, Berlin, Germany

Protein purification buffers

PBS/ pH 7,4	137 mM NaCl 2.7 mM KCl 8 mM Na_2HPO_4 1.5 mM KH_2PO_4
DHFR Lysis buffer A/ pH 7,4	300 mM NaCl 50 mM NaH_2PO_4 20 mM Imidazol
Wash buffer A/ pH 8	300 mM NaCl 50 mM NaH_2PO_4 30 mM Imidazol

Elution buffer A/ pH 8	300 mM NaCl 50 mM NaH_2PO_4 250 mM Imidazol
Av17	
Lysis buffer B/ pH 8	PBS
Wash buffer B/ pH6	PBS 10 % (v/v) Glycerin
Elution buffer B/ pH 4	PBS 10 % (v/v) Glycerin
Neutralisation buffer/ pH9	1 M Tris
mAv17	
Lysis buffer/ pH 8	PBS/0.1 % (v/v) Triton
Wash buffer/ pH 5.8	PBS/10 % (v/v) Glycerin
Elution buffer/ pH 3	PBS/10 % (v/v) Glycerin
Neutralisation buffer/ pH10	PBS
Cysele2	
Lysis buffer/ pH8	8 M Urea 100 mM NaH_2PO_4 10 mM Tris
Wash buffers/ pH6.3 and pH5.9	see lysis buffer
Elution buffer/ pH4.5	see lysis buffer
Dialysis 1/ pH7.4	1 M Urea 0.1 M NaH_2PO_4 10 mM Tris
Dialysis 2/ pH7.4	PBS
Dialysis 3/ pH7.4	PBS/0.05 % (v/v) Triton
LB-medium	10 g Trypton 10 g NaCl 5 g yeast extract ad. 1 l H_2O dest.
Activity test reaction buffer	0.2 M K_2HPO_4 0.2 M KH_2PO_4 0.004 M DTT 0.004 M EDTA 0.01 % (v/v) Brij
BAL-solution	PBS (Dulbeccos w/o) Complete mini protease inhibitor

6.7.3. Chemicals, biologicals and recombinant cytokines

^{3}H Thymidin	ICN, Costa Mesa, USA
BSA, fraction V	AppliChem, Darmstadt, Germany
BAPNA (92µM)	Bachem, Weil a. Rhein, Germany
Proteinase inhibitor cocktail	Roche, Mannheim, Germany
Paraformaldehyde	Sigma, München, Germany
Papain (100µM)	Sigma, München, Germany

6.7.4. Commercial Kits

Foxp3 detection kit (clone FJK-16s, PE-coupled)	eBiosciences, San Diego, CA
OptEIA ELISA kits	BD Biosciences, Heidelberg, Germany
TGF-β ELISA-set	R&D Systems, Minneapolis, MN
BCA kit	Pierce, USA
TaqMan primer and probes	Applied Biosystems, Darmstadt, Germany
TaqMan cDNA synthesis kit	Applied Biosystems, Germany
TaqMan PCR master mix	Applied Biosystems, Germany
RNeasy Mini kit	Qiagen, Hilden, Germany
QIAshredder spin columns	Qiagen, Hilden, Germany
LAL QCL-1000	Cambrex, USA
Cytokine bead array 13-Plex	R&D systems, Minneapolis, USA
Profos endotrap system red	Profos, Regensburg, Germany

6.7.5. Antibodies and secondary reagents

Specificity	fluorochrome	clone	Purchased from
αCD3	PE	145.2C11	DRFZ*
αCD4	FITC/PerCP	RM4-5	BD Biosciences
αCD19	FITC	ID3	DRFZ*
αCD25	APC/PerCP-Cy5.5	PC61	BD Biosciences
αCD103	PE	M290	BD Biosciences
αCD103	- (biotin)	M290	DRFZ*
αF4/80	Cy5	F4/80	DRFZ*
αIL10R	-	1B1	Falk, Rötschke**
FcγR	-	2.4G2	DRFZ*
Rat IgG	-	polyclonal	Dianova
Secondary reagents			
SA-PE-Cy7	PECy7		BD Biosciences
SA-PE	PE		BD Biosciences

*   mAb was a kind gift of the German Arthritis Research Centre.
** mAb was a kind gift of Dr. K. Falk and Dr. O. Rötschke, MDC Buch

6.7.6. Software
FlowJo (Tristar)

BD FACSDiva (BD Biosciences)

Prism (GraphPad)

7. Abbreviations

Ab	antibody
A.D.	atopic dermatitis
Ag	antigen
APC	allophycocyanin
APC	antigen presenting cell
AHR	airway hyperreactivity
Av17	*Acanthocheilonema viteae* 17kDa immunomodulator (cystatin)
BAL	bronchoalveolar lavage
BALF	bronchoalveolar lavage fluid
Bio	biotin
BSA	bovine serum albumin
CD	cluster of differentiation
DC	dendritic cell
DHFR	dihydrofolat reductase
e.c.	epicutanous
ELISA	enzyme-linked immunosorbent assay
EU	endotoxin unit
FACS	fluorescence activated cell sorter
FCS	fetal calf serum
FITC	fluorescein isothiocyanate
Foxp3	forkhead box transcription factor p3
FSC	forward scatter
G	G force
GM-CSF	granulocyte-macrophage colony stimulating factor
HDM	house dust mite
Hp	*Heligmosomoides polygyrus*
h	hour
IFN	interferon
Ig	immune globuline
IL	interleukin
IL-10 R	interleukin-10 receptor
ILN	inguinal lymph node

Abbreviations

ILNC	inguinal lymph node cell
i.n.	intranasal
i.p.	intraperitoneal
KO	knock out
LU	lab unit
µg	microgram
µl	microliter
mg	milligram
ml	milliliter
M	molar
mAb	monoclonal antibody
MHC	major histocompatibility complex
min	minute
MLN	mesenteric lymph node
MLNC	mesenteric lymph node cell
MLV	multilamellar vesicle
ng	nanogram
NK	natural killer cell
NO	nitric oxide
OVA	ovalbumin
PBLN	peribronchial lymph node
PBLNC	peribronchial lymph node cell
PBS	phosphate buffered saline
PCR	polymerase chain reaction
PE	phycoerythrin
PEC	peritoneal excudate cells
Penh	pause enhanced
pg	picogram
RT	room temperature
SA	streptavidin
Treg	regulatory T cell
Th1	T helper cell 1
Th2	T helper cell 2

TGF	transforming growth factor
TNF	tumor necrosis factor

8. References

. "en.wikipedia.org/wiki/clodronate."
. "en.wikipedia.org/wiki/Cysteine_protease."
. "www.clodronateliposomes.org."
Abrahamson, M. (1994). "Cystatins." <u>Meth Enzymol</u> **244**: 685-700.
Adachi, H., and M. Tsujimoto (2006). "Endothelial scavenger receptors." <u>Prog. Lipid Res.</u> **45**: 379–404.
Akbari, O., R. H. DeKruyff, and D. T. Umetsu (2001). "Pulmonary dendritic cells producing IL-10 mediate tolerance induced by respiratory exposure to antigen." <u>Nat.Immunol</u>(2): 725–731.
Akdis, C. A. (2006). "Allergy and hypersensitivity: Mechanisms of allergic disease." <u>Current Opinion in Immunology</u> **18**(6): 718-726.
Aksoy, E., Zouain, C.S., Vanhoutte, F., Fontaine, J., Pavelka, N., Thieblemont, N., Willems, F., Ricciardi-Castagnoli, P., Goldman, M., Capron, M., Ryffel, B., Trottein, F. (2005). "Double-stranded RNAs from the helminth parasite Schistosoma activate TLR3 in dendritic cells." <u>J Biol Chem</u> **280**(1): 277-83.
Alcorn, J., Rinaldi, LM, Jaffe, EF, van Loon, M, Bates, JHT, Janssen-Heininger, YMW, Irvin, CG. (2007). "Transforming Growth Factor-beta1 Suppresses Airway Hyperresponsiveness in Allergic Airway Disease." <u>Am J Respir Crit Care Med</u> **176**(10): 974-982.
Allen, J., MacDonald, AS (1998). "Profound suppression of cellular proliferation mediated by the secretions of nematodes." <u>Parasite Immunol</u> (20): 241–247.
Anderson, C. F., Gerber, J. S., Mosser, D. M. (2002a). "Modulating macrophage function with IgG immune complexes." <u>J Endotoxin Res</u> **8**(6): 477-81.
Anderson, C. F. and D. M. Mosser (2002b). "Cutting edge: biasing immune responses by directing antigen to macrophage Fc gamma receptors." <u>J Immunol</u> **168**(8): 3697-701.
Anderson, C. F. and D. M. Mosser (2002c). "A novel phenotype for an activated macrophage: the type 2 activated macrophage." <u>J Leukoc Biol</u> **72**(1): 101-106.
Anthony, R. M., Rutitzky LI, Urban JF Jr, Stadecker MJ, Gause WC (2007). "Protective immune mechanisms in helminth infection." <u>Nat Rev Immunol.</u> **7**(12): 975-87.
Anthony, R. M., Urban, J. F., Alem, F., Hamed, H. A., Rozo, C.T., Boucher, J., Van Rooijen, N., Gause, W.C. (2006). "Memory TH2 cells induce alternatively activated macrophages to mediate protection against nematode parasites." <u>Nat Med</u> **12**(8): 955-960.
Araujo, M., de Carvalho, E. (2006). "Human Schistosomiasis Decreases Immune Responses to Allergens and Clinical Manifestations of Asthma." <u>Chem Immunol Allergy</u> **90**: 29-44
Araujo, M., Hoppe, B., Medeiros, Jr M., Alcantara, L., Almeida, MC., Schriefer, A., Oliveira, RR., Kruschewsky, R., Figueiredo, JP., Cruz, AA., Carvalho, EM. (2004). "Impaired T Helper 2 Response to Aeroallergen in Helminth-Infected Patients with Asthma." <u>J Infect Dis</u> **190**(10): 1797-1803.
Asadullah, K., Sterry, W., Volk, H. D. (2003). "Interleukin-10 Therapy--Review of a New Approach." <u>Pharmacol Rev</u> **55**(2): 241-269.
Atochina, O., Daly-Engel, T, Piskorska, D, McGuire, E, Harn, DA. (2001). "A Schistosome-Expressed Immunomodulatory Glycoconjugate Expands Peritoneal Gr1+ Macrophages That Suppress Naive CD4+ T Cell Proliferation Via an IFN-gamma and Nitric Oxide-Dependent Mechanism." <u>J Immunol</u> **167**(8): 4293-4302.

Auriault C, O. M., Torpier G, Eisen H, Capron A (1981). "Proteolytic cleavage of IgG bound to the Fc receptor of Schistosoma mansoni schistosomula." Parasite Immunol **3**(1): 33-44.

Averbeck, M., Gebhardt, C., Emmrich, F., Treudler, R., Simon, J. C. (2007). "Immunologic principles of allergic disease." J Dtsch Dermatol Ges **5**(11): 1015-28.

Bach, J.-F. (2002). "The Effect of Infections on Susceptibility to Autoimmune and Allergic Diseases." N Engl J Med **347**(12): 911-920.

Barlan, I. B., Bahceciler, N., Akdis, M., Akdis, C.A. (2006). "Bacillus Calmette-Guerin, Mycobacterium bovis, as an immunomodulator in atopic diseases." Immunol Allergy Clin North Am **26**(2): 365-77.

Barrett, A. (1986). "The cystatins: a diverse superfamily of cysteine peptidase inhibitors." Biomed Biochim Acta **45** 1363–1374.

Bashir, M., Andersen P, Fuss IJ, Shi HN, Nagler-Anderson C (2002). "An enteric helminth infection protects against an allergic response to dietary antigen." J Immunol **169**(6): 3284-92.

Batten, M., Li J, Yi S, Kljavin NM, Danilenko DM, Lucas S, Lee J, de Sauvage FJ, Ghilardi N. (2006). " Interleukin 27 limits autoimmune encephalomyelitis by suppressing the development of interleukin 17-producing T cells." Nat Immunol. **7**(9): 929-36.

Baumgart, M., F. Tompkins, J. Leng, and M. Hesse (2006). "Naturally occurring CD4+Foxp3+ regulatory T cells are an essential, IL-10-independent part of the immunoregulatory network in Schistosoma mansoni egg-induced inflammation." J. Immunol(176): 5374-5387.

Beier, K. C., A. Hutloff, M. Löhning, T. Kallinich, R. A. Kroczek, E. Hamelmann. (2004). "Inducible costimulator-positive T cells are required for allergen-induced local B-cell infiltration and antigen-specific IgE production in lung tissue. ." J. Allergy Clin. Immunol. **114**: 775–782.

Bilsborough, J., Leung DY, Maurer M, Howell M, Boguniewicz M, Yao L, Storey H, LeCiel C, Harder B, Gross JA. (2006). "IL-31 is associated with cutaneous lymphocyte antigen-positive skin homing T cells in patients with atopic dermatitis." J Allergy Clin Immunol **117**(2): 418-25.

Blackburn, C. C., Selkirk, M.E. (1992). "Inactivation of platelet-activating factor by a putative acetylhydrolase from the gastrointestinal nematode parasite Nippostrongylus brasiliensis." Immunology **75**(1): 41-6.

Blümchen, K., Gerhold, K., Thorade, I., Seib, C., Wahn, U., Hamelmann, E. (2004). "Oral administration of desloratadine prior to sensitization prevents allergen-induced airway inflammation and hyper-reactivity in mice." Clin Exp Allergy **34**(7): 1124-30.

Blumchen, K., K. Gerhold, M. Schwede, B. Niggemann, A. Avagyan, A. M. Dittrich, B. Wagner, H. Breiteneder, E. Hamelmann. (2006). "Effects of established allergen sensitization on immune and airway responses after secondary allergen sensitization. ." J. Allergy Clin. Immunol **118**: 615–621.

Boguniewicz, M. (2004). "Update on atopic dermatitis: insights into pathogenesis and new treatment paradigms." Allergy Asthma Proc. **25**(5): 279-82.

Bohle, B., Kinaciyan, T., Gerstmayr, M., Radakovics, A., Jahn-Schmid, B., Ebner, C. (2007). "Sublingual immunotherapy induces IL-10-producing T regulatory cells, allergen-specific T-cell tolerance, and immune deviation." J Allergy Clin Immunol. **120**(3): 707-13.

Boitelle, A., Di Lorenzo C, Scales HE, Devaney E, Kennedy MW, Garside P, Lawrence CE. (2005). "Contrasting effects of acute and chronic gastro-intestinal helminth infections on a heterologous immune response in a transgenic adoptive transfer model." Int J Parasitol **35**(7): 765-75.

Bolton, S. J., McNulty, C. A., Thomas, R. J., Hewitt, C. R., Wardlaw, A. J. (2003). "Expression of and functional responses to protease-activated receptors on human eosinophils." J Leukoc Biol **74**(1): 60-8.

Bradding, P., A. F. Walls, and S. T. Holgate (2006). "The role of the mast cell in the pathophysiology of asthma." J. Allergy Clin. Immunol **117**: 1277-1284.

Braman, S. S. (2006). "The Global Burden of Asthma." Chest **130**(1_suppl): 4S-12.

Brattig, N. (2004). "Pathogenesis and host responses in human onchocerciasis: impact of Onchocerca filariae and Wolbachia endobacteria." Microbes Infect **6**(1): 113-28.

Braun-Fahrländer, C., Riedler, J., Herz, U., Eder, W., Waser, M., Grize, L., Maisch, S., Carr, D., Gerlach, F., Bufe, A., Lauener, RP., Schierl, R., Renz, H., Nowak, D., von Mutius, E.; Allergy and Endotoxin Study Team (2002). "Environmental exposure to endotoxin and its relation to asthma in school-age children." N Engl J Med **347**(12): 869-77.

Brem-Exner, B. G., Sattler, C., Hutchinson, J. A., Koehl, G. E., Kronenberg, K., Farkas, S., Inoue, S., Blank, C., Knechtle, S. J., Schlitt, H. J., Fandrich, F., Geissler, E. K. (2008). "Macrophages driven to a novel state of activation have anti-inflammatory properties in mice." J Immunol **180**(1): 335-49.

Brys, L., Beschin, A, Raes, G, Ghassabeh, GH, Noel, W, Brandt, J, Brombacher, F, Baetselier, PD. (2005). "Reactive Oxygen Species and 12/15-Lipoxygenase Contribute to the Antiproliferative Capacity of Alternatively Activated Myeloid Cells Elicited during Helminth Infection." J Immunol **174**(10): 6095-6104.

Buer, J., Lanoue, A., Franzke, A., Garcia, C., von Boehmer, H., Sarukhan, A. (1998). "Interleukin 10 secretion and impaired effector function of major histocompatibility complex class II–restricted T cells anergized in vivo." J Exp Med **187**: 177–83.

Chen, C.-C., S. Louie, et al. (2005). "Concurrent Infection with an Intestinal Helminth Parasite Impairs Host Resistance to Enteric Citrobacter rodentium and Enhances Citrobacter-Induced Colitis in Mice." Infect Immun **73**(9): 5468-5481.

Chen, L., Martinez O, Overbergh L, Mathieu C, Prabhakar BS, Chan LS. (2004). "Early up-regulation of Th2 cytokines and late surge of Th1 cytokines in an atopic dermatitis model." Clin Exp Immunol **138**(3): 375-387.

Chen, Y., Kuchroo, V.K., Inobe, J., Hafler, D.A., Weiner, H.L. (1994). "Regulatory T cell clones induced by oral tolerance: suppression of autoimmune encephalomyelitis." Science **265**: 1237–40.

Chialda, L., M. Zhang, K. Brune, and A. Pahl. (2005). "Inhibitors of mitogenactivated protein kinases differentially regulate costimulated T cell cytokine production and mouse airway eosinophilia." Respir. Res **6**: 36–54.

Cooke, A. (2006). "Th17 Cells in Inflammatory Conditions." Rev Diabet Stud. **3**(2): 72–75.

Cooper, P. (2002). "Can intestinal helminth infections (geohelminths) affect the development and expression of asthma and allergic disease?" Clin Exp Immunol **128**(3): 398-404.

Cooper, P. (2004). "The potential impact of early exposures to geohelminth infections on the development of atopy." Clin Rev Allergy Immunol **26**(1): 5-14.

Cooper, P. J., Barreto, M. L., Rodrigues, L. C. (2006). "Human allergy and geohelminth infections: a review of the literature and a proposed conceptual model to guide the investigation of possible causal associations." Br Med Bull **79-80**(1): 203-218.

Culley, F., Brown, A., Conroy, DM., Sabroe, I., Pritchard, DI., Williams TJ. (2000). "Eotaxin is specifically cleaved by hookworm metalloproteases preventing its action in vitro and in vivo." J Immunol **165**(11): 6447-53.

Dagoye, D., Z. Bekele, et al. (2003). "Wheezing, Allergy, and Parasite Infection in Children in Urban and Rural Ethiopia." Am J Respir Crit Care Med **167**(10): 1369-1373.

Dahten, A., Koch, C., Ernst, D., Schnöller, C., Hartmann, S., Worm, M. (2008). "Systemic PPAR[gamma] Ligation Inhibits Allergic Immune Response in the Skin." J Invest Dermatol Epub ahead of print.

Dainichi, T., Maekawa, Y., Ishii, K., Zhang, T., Nashed, B. F., Sakai, T., Takashima, M., Himeno, K. (2001). "Nippocystatin, a Cysteine Protease Inhibitor from Nippostrongylus brasiliensis, Inhibits Antigen Processing and Modulates Antigen-Specific Immune Response." Infect Immun **69**(12): 7380-7386.

de Macedo Soares, M. F., de Macedo, M.S. (2007). "Modulation of anaphylaxis by helminth-derived products in animal models." Curr Allergy Asthma Rep **7**(1): 56-61.

Delayre-Orthez, C., de Blay, F., Frossard, N,. Pons, F. (2004). "Dose-dependent effects of endotoxins on allergen sensitization and challenge in the mouse." Clin Exp Allergy **34**(11): 1789-95.

Dittrich, A. M., Erbacher A, Specht S, Diesner F, Krokowski M, Avagyan A, Stock P, Ahrens B, Hoffmann WH, Hoerauf A, Hamelmann E. (2008). "Helminth infection with Litomosoides sigmodontis induces regulatory T cells and inhibits allergic sensitization, airway inflammation, and hyperreactivity in a murine asthma model." J Immunol **180**(3): 1792-9.

Dugina, T. N., Kiseleva, E. V., Glusa, E., Strukova, S. M. (2003). "Activation of mast cells induced by agonists of proteinase-activated receptors under normal conditions and during acute inflammation in rats." Eur J Pharmacol **471**(2): 141-7.

Dunne, D., Cooke A (2005). "A worm's eye view of the immune system: consequences for evolution of human autoimmune disease." Nat Rev Immunol **5**(5): 420-6.

Edwan, J. H., G. Perry, J. E. Talmadge, D. K. Agrawal. (2004). "Flt-3 ligand reverses late allergic response and airway hyper-responsiveness in a mouse model of allergic inflammation." J. Immunol **172**: 5016–5023.

Edwards, J., Zhang, X, Frauwirth, KA, Mosser, DM. (2006). "Biochemical and functional characterization of three activated macrophage populations." J Leukoc Biol. **80**(6): 1298-1307.

Effros, R. M., Nagaraj, H. (2007). "Asthma: new developments concerning immune mechanisms, diagnosis and treatment." Curr Opin Pulm Med **13**(1): 37-43.

Elliott, D., Setiawan T, Metwali A, Blum A, Urban JF Jr, Weinstock JV. (2004). "Heligmosomoides polygyrus inhibits established colitis in IL-10-deficient mice." Eur J Immunol. **34**(10): 2690-8.

Fallon, P. G. and N. E. Mangan (2007). "Suppression of TH2-type allergic reactions by helminth infection." Nat Rev Immunol **7**(3): 220-30.

Fattouh, R., Midence, NG, Arias, K, Johnson, JR, Walker, TD, Goncharova, S, Souza, KP, Gregory, RC, Jr, Lonning, S, Gauldie, J, Jordana, M. (2008). "Transforming Growth Factor-beta Regulates House Dust Mite-induced Allergic Airway Inflammation but Not Airway Remodeling." Am J Respir Crit Care Med **177**(6): 593-603.

Fehervari, Z., and S. Sakaguchi (2004). "CD4+ Tregs and immune control." J. Clin.Invest **114**: 1209–1217.

Finkelman, FD, Shea.-Donohue T, Morris SC, Gildea L, Strait R, Madden KB, Schopf L, Urban JF Jr (2004). "Interleukin-4- and interleukin-13-mediated host protection against intestinal nematode parasites." Immunol Rev **201**: 139-55.

Fiset, P. O., Leung, D.Y., Hamid, Q. (2006). "Immunopathology of atopic dermatitis." J Allergy Clin Immunol. **118**(1): 287-90.

Fort, M. M., Cheung, J., Yen, D., Li, J., Zurawski, S.M., Lo, S., Menon, S., Clifford, T., Hunte, B., Lesley, R., Muchamuel, T., Hurst, S.D., Zurawski, G., Leach, M.W., Gorman, D.M., Rennick, D.M. (2001). "IL-25 induces IL-4, IL-5, and IL-13 and Th2-associated pathologies in vivo." Immunity **15**(6): 985-95.

Fox, J. G., P. Beck, et al. (2000). "Concurrent enteric helminth infection modulates inflammation and gastric immune responses and reduces helicobacter-induced gastric atrophy." Nat Med **6**(5): 536-542.

Frankenberger, M, Häussinger K, Ziegler-Heitbrock L (2005). "Liposomal methylprednisolone differentially regulates the expression of TNF and IL-10 in human alveolar macrophages " Int Immunopharmacol **5**(2): 289-99.

Frohlich, A., Marsland, BJ, Sonderegger, I, Kurrer, M, Hodge, MR, Harris, NL, Kopf, M. (2007). "IL-21 receptor signaling is integral to the development of Th2 effector responses in vivo." Blood **109**(5): 2023-2031.

Fu, C. L., Y. L. Ye, Y. L. Lee, and B. L. Chiang (2006). "Effects of overexpression of IL-10, IL-12, TGF-beta and IL-4 on allergen induced change in bronchial responsiveness." Respir. Res **7**: 72-85.

Fukaura, H., Kent, S.C., Pietrusewicz, M.J., Khoury, S.J., Weiner, H.L., Hafler, D.A. (1996). "Induction of circulating myelin basic protein and proteolipid protein-specific transforming growth factor–b1–secreting Th3 T cells by oral administration of myelin in multiple sclerosis patients." J Clin Invest **98**: 70-7.

Furze, R. C., T. Hussell, and M. E. Selkirk (2006). "Amelioration of influenza-induced pathology in mice by coinfection with Trichinella spiralis." Infect. Immun **74**: 1924-1932.

Gajewski, T. F. and F. W. Fitch (1988). "Anti-proliferative effect of IFN-gamma in immune regulation. I. IFN- gamma inhibits the proliferation of Th2 but not Th1 murine helper T lymphocyte clones." J Immunol **140**(12): 4245-4252.

Gause, W. C., Urban JF Jr, Stadecker MJ. (2003). "The immune response to parasitic helminths: insights from murine models." Trends Immunol. **24**(5): 269-77.

Gerber, J., Mosser, DM. (2001). "Reversing Lipopolysaccharide Toxicity by Ligating the Macrophage Fc-gamma- Receptors." J Immunol **166**(11): 6861-6868.

Gerhold, K., A. Avagyan, C. Seib, R. Frei, J. Steinle, B. Ahrens, A. M. Dittrich, K. Blumchen, R. Lauener, and E. Hamelmann (2006). "Prenatal initiation of endotoxin airway exposure prevents subsequent allergen-induced sensitization and airway inflammation in mice." J. Allergy Clin. Immunol. **118**: 666-673.

Gerhold, K., Blümchen, K., Bock, A., Seib, C., Stock, P., Kallinich, T., Löhning, M., Wahn, U., Hamelmann, E. (2002). "Endotoxins prevent murine IgE production, T(H)2 immune responses, and development of airway eosinophilia but not airway hyperreactivity." J Allergy Clin Immunol **110**(1): 110-6.

Godfraind, C., Louahed, J., Faulkner, H., Vink, A., Warnier, G., Grencis, R., Renauld, J. (1998). "Intraepithelial Infiltration by Mast Cells with Both Connective Tissue-Type and Mucosal-Type Characteristics in Gut, Trachea, and Kidneys of IL-9 Transgenic Mice." J Immunol **160**(8): 3989-3996.

Goerdt, S., Politz, O., Schledzewski, K., Birk, R., Gratchev, A., Guillot, P., Hakiy, N., Klemke, C.D., Dippel, E., Kodelja, V., Orfanos, C.E. (1999). "Alternative versus classical activation of macrophages." Pathobiology **67**(5-6): 222-6.

Gomez-Escobar N, L. E., Maizels RM. (1998). "A novel member of the transforming growth factor-beta (TGF-beta) superfamily from the filarial nematodes Brugia malayi and B. pahangi." Exp Parasitol **88**(3): 200-9.

Goodridge, H. S., Deehan, M.R., Harnett, W., Harnett, M.M. (2005b). "Subversion of immunological signalling by a filarial nematode phosphorylcholine-containing secreted product." Cell Signal. **17**(1): 11-6.

Goodridge, H. S., Marshall, F.A., Else, K.J., Houston, K.M., Egan, C., Al-Riyami, L., Liew, F.Y., Harnett, W., Harnett, M.M. (2005a). " Immunomodulation via novel use of TLR4 by the filarial nematode phosphorylcholine-containing secreted product, ES-62." J Immunol **174**(1): 284-93.

Goodridge, H. S., Marshall, F.A., Wilson, E.H., Houston, K.M., Liew, F.Y., Harnett, M.M., Harnett, W. (2004). "In vivo exposure of murine dendritic cell and macrophage bone marrow progenitors to the phosphorylcholine-containing filarial nematode glycoprotein ES-62 polarizes their differentiation to an anti-inflammatory phenotype." Immunology 113(4): 491-8.

Goodridge, H. S., Wilson, E.H., Harnett, W., Campbell, C.C., Harnett, M.M., Liew, F. Y. (2001). "Modulation of Macrophage Cytokine Production by ES-62, a Secreted Product of the Filarial Nematode Acanthocheilonema viteae." J Immunol 167(2): 940-945.

Gordon, S. (2003). "Alternative activation of macrophages." Nat Rev Immunol 3(1): 23-35.

Groux, H., O'Garra, A., Bigler, M., Rouleau, M., Antonenko, S., de Vries, J.E., Roncarolo, M.G. (1997). "A CD4+ T-cell subset inhibits antigen-specific T-cell responses and prevents colitis." Nature 389: 737–42.

Hafler, D. A., Kent, S.C., Pietrusewicz, M.J., Khoury, S.J., Weiner, H.L., Fukaura, H. (1997). "Oral administration of myelin induces antigen-specific TGF-b1 secreting T cells in patients with multiple sclerosis." Ann NY Acad Sci 835: 120–31.

Hagen, J. (2007). Einfluss rekombinanter Nematodenantigene auf die Reifung dendritischer Zellen und die Polarisierung der CD4+-T-Zell-Antwort. Lehrstuhl für Molekulare Parasitologie. Berlin, Humboldt-Universität zu Berlin.

Haileamlak, A., Dagoye D, Williams H, Venn AJ, Hubbard R, Britton J, Lewis SA (2005). "Early life risk factors for atopic dermatitis in Ethiopian children." J Allergy Clin Immunol 115(2): 370-6.

Hamelmann, E., Beyer, K., Gruber, C., Lau, S., Matricardi, P.M., Nickel, R., Niggemann, B., Wahn, U. (2008). "Primary prevention of allergy: avoiding risk or providing protection?" Clin Exp Allergy 38(2): 233-45.

Hamelmann, E., Gelfand, E.W. (2001). "IL-5-induced airway eosinophilia--the key to asthma?" Immunol Rev 179: 182-91.

Hamelmann, E., Oshiba, A., Schwarze, J., Bradley, K., Loader, J., Larsen, G., Potter, T., Gelfand, E.W. (1997a). "Allergen-Specific IgE and IL-5 are essential for the development of airway hyper-responsiveness." Am. J. Resp. Cell Mol. Biol. 16(6): 674-682

Hamelmann, E., Schwarze J, Takeda K, Oshiba A, Larsen GL, Irvin CG, Gelfand EW. (1997c). "Noninvasive measurement of airway responsiveness in allergic mice using barometric plethysmography." Am. J. Respir. Crit. Care. Med 156(3 Pt 1): 766-775.

Hamelmann, E., Vella, A.T., Oshiba, A., Kappler, J.W., Marrack, P., Gelfand, E.W. (1997b). "Allergic airway sensitization induces T cell activation but not airway hyperresponsiveness in B cell-deficient mice." Proc Natl Acad Sci U S A 94(4): 1350-1355.

Harnett, W. and M. M. Harnett (2006). "Filarial nematode secreted product ES-62 is an anti-inflammatory agent: therapeutic potential of small molecule derivatives and ES-62 peptide mimetics " Clin Exp Pharmacol Physiol. 33(5-6): 511-518.

Harris, J., Mason DE, Li J, Burdick KW, Backes BJ, Chen T, Shipway A, Van Heeke G, Gough L, Ghaemmaghami A, Shakib F, Debaene F, Winssinger N. (2004). "Activity profile of dust mite allergen extract using substrate libraries and functional proteomic microarrays." Chem Biol. 11(10): 1361-72.

Hartl, D., Koller, B., Mehlhorn, A.T., Reinhardt, D., Nicolai, T., Schendel, D.J., Griese, M., Krauss-Etschmann, S. (2007). "Quantitative and functional impairment of pulmonary CD4+CD25hi regulatory T cells in pediatric asthma." J Allergy Clin Immunol 119(5): 1258-1266.

Hartmann, S., Kyewski B, Sonnenburg B, Lucius R (1997). "A filarial cysteine protease inhibitor down-regulates T cell proliferation and enhances interleukin-10 production." Eur J Immunol 27(9): 2253-60.

Hartmann, S., Lucius, R. (2003). "Modulation of host immune responses by nematode cystatins." Int J Parasitol 33(11): 1291-302.

Hartmann, S., Schonemeyer, A, Sonnenburg, B, Vray, B, Lucius, R. (2002). "Cystatins of filarial nematodes up-regulate the nitric oxide production of interferon-gamma-activated murine macrophages." Parasite Immunol. 24(5): 253-262.

Hawrylowicz, C. M. (2005a). "Regulatory T cells and IL-10 in allergic inflammation." J. Exp. Med. 202: 1459-1463.

Hawrylowicz, C. M., O`Garra, A. (2005b). "Potential role of interleukin-10-secreting regulatory T cells in allergy and asthma." Nat Immunol 5: 271-283.

He, J., Song Y, Ueyama N, Harada A, Azakami H, Kato A (2005). "Characterization of recombinant amyloidogenic chicken cystatin mutant I66Q expressed in yeast." J Biochem. 137(4): 477-85.

Henskens, Y., Veerman EC, Nieuw Amerongen AV (1996). "Cystatins in health and disease." Biol Chem Hoppe Seyler 377(2): 71-86.

Herbert, D., Hölscher C, Mohrs M, Arendse B, Schwegmann A, Radwanska M, Leeto M, Kirsch R, Hall P, Mossmann H, Claussen B, Förster I, Brombacher F (2004). "Alternative macrophage activation is essential for survival during schistosomiasis and downmodulates T helper 1 responses and immunopathology." Immunity 20(5): 623-35.

Hesse, M., Modolell, M, La Flamme, AC, Schito, M, Fuentes, JM, Cheever, AW, Pearce, EJ, Wynn, TA. (2001). "Differential Regulation of Nitric Oxide Synthase-2 and Arginase-1 by Type 1/Type 2 Cytokines In Vivo: Granulomatous Pathology Is Shaped by the Pattern of L-Arginine Metabolism." J Immunol 167(11): 6533-6544.

Hirahara, K., Liu L, Clark RA, Yamanaka K, Fuhlbrigge RC, Kupper TS. (2006). "The majority of human peripheral blood CD4+CD25highFoxp3+ regulatory T cells bear functional skin-homing receptors. ." J Immunol. 177(7): 4488-94.

Hirata, M., Hirata K, Hara T, Kawabuchi M, Fukuma T. (2005). "Expression of TGF-beta-like molecules in the life cycle of Schistosoma japonicum." Parasitol Res 95(6): 367-73.

Holt, P. G., J. E. Batty, and K. J. Turner (1981). "Inhibition of specific IgE responses in mice by pre-exposure to inhaled antigen." Immunology(42): 409–417.

Hoves, S., Krause, SW, Schutz, C, Halbritter, D, Scholmerich, J, Herfarth, H, Fleck, M. (2006). "Monocyte-Derived Human Macrophages Mediate Anergy in Allogeneic T Cells and Induce Regulatory T Cells." J Immunol 177(4): 2691-2698.

Huehn, J., Siegmund K, Lehmann JC, Siewert C, Haubold U, Feuerer M, Debes GF, Lauber J, Frey O, Przybylski GK, Niesner U, de la Rosa M, Schmidt CA, Bräuer R, Buer J, Scheffold A, Hamann A. (2004). "Developmental stage, phenotype, and migration distinguish naive- and effector/memory-like CD4+ regulatory T cells." J Exp Med. 199(3): 303-13.

Hurst, S. D., Muchamuel T, Gorman DM, Gilbert JM, Clifford T, Kwan S, Menon S, Seymour B, Jackson C, Kung TT, Brieland JK, Zurawski SM, Chapman RW, Zurawski G, Coffman RL. (2002). "New IL-17 family members promote Th1 or Th2 responses in the lung: in vivo function of the novel cytokine IL-25." J Immunol. 169(1): 443-53.

Iellem, A., Colantonio L, D'Ambrosio D. (2003). "Skin-versus gut-skewed homing receptor expression and intrinsic CCR4 expression on human peripheral blood CD4+CD25+ suppressor T cells." Eur J Immunol 33(6): 1488-96.

Ikeda, K., Nakajima, H., Suzuki, K., Kagami, S., Hirose, K., Suto, A., Saito, Y., Iwamoto, I. (2003). "Mast cells produce interleukin-25 upon Fcepsilon RI-mediated activation." Blood 101(9): 3594-3596.

ISAAC, (The international study of asthma and allergies in childhood steering committee) (1998). "Worldwide variation in prevalence of symptoms of asthma, allergic rhino-conjunctivitis, and atopic eczema: ISAAC." Lancet(351): 1225-1232.

Itami, D. M., Oshiro, T.M., Araujo, C.A., Perini, A., Martins, M.A., Macedo, M.S., Macedo-Soares, M.F. (2005). "Modulation of murine experimental asthma by Ascaris suum components." Clin Exp Allergy 35(7): 873-9.

Ito, S., Ansari, P, Sakatsume, M, Dickensheets, H, Vazquez, N, Donnelly, RP, Larner, AC, Finbloom, DS. (1999). "Interleukin-10 Inhibits Expression of Both Interferon alpha - and Interferon gamma - Induced Genes by Suppressing Tyrosine Phosphorylation of STAT1." Blood 93(5): 1456-1463.

Jankovic, D., Kullberg, MC, Feng, CG, Goldszmid, RS, Collazo, CM, Wilson, M, Wynn, TA., Kamanaka, M, Flavell, RA, Sher, A. (2007). "Conventional T-bet+Foxp3- Th1 cells are the major source of host-protective regulatory IL-10 during intracellular protozoan infection." J Exp Med 204(2): 273-283.

Janowski, R., Kozak, M, Jankowska, E, Grzonka, Z, Grubb, A, Abrahamson, M, Jaskolski, M. (2001). "Human cystatin C, an amyloidogenic protein, dimerizes through three-dimensional domain swapping." Nat Struct Mol Biol 8(4): 316-320.

John, R. J., Rusznak C, Ramjee M, Lamont AG, Abrahamson M, Hewitt EL (2000). "Functional effects of the inhibition of the cysteine protease activity of the major house dust mite allergen Der p 1 by a novel peptide-based inhibitor." Clin Exp Allergy. 30(6): 784-793.

Kato, T., Takai T, Mitsuishi K, Okumura K, Ogawa H. (2005). "Cystatin A inhibits IL-8 production by keratinocytes stimulated with Der p 1 and Der f 1: biochemical skin barrier against mite cysteine proteases." J Allergy Clin Immunol 116(1): 169-76.

Kay, A. B. (2001). "Allergy and allergic disease." N.Engl.J.Med 344: 30-37.

Kennedy, M. W. (2000). "Immune response to Anisakis simplex and other ascarid nematodes." Allergy 55(s59): 7-13.

Kenyon, N. J., Ward, R. W., McGrew, G., Last, J. A. (2003). "TGF-beta1 causes airway fibrosis and increased collagen I and III mRNA in mice." Thorax 58(9): 772-777.

King, C. L., M. Connelly, et al. (2001). "Transmission Intensity Determines Lymphocyte Responsiveness and Cytokine Bias in Human Lymphatic Filariasis." J Immunol 166(12): 7427-7436.

Kitagaki, K., T. R. Businga, D. Racila, D. E. Elliott, J. V. Weinstock, and J. N. Kline (2006). "Intestinal helminths protect in a murine model of asthma." J. Immunol 177: 1628-1635.

Koga, C., Kabashima, K, Shiraishi, N, Kobayashi, M, Tokura, Y. (2008). "Possible Pathogenic Role of Th17 Cells for Atopic Dermatitis." J Invest Dermatol.

Kohm, A., McMahon, JS, Podojil, JR, Begolka, WS, DeGutes, M, Kasprowicz, DJ, Ziegler, SF, Miller, SD. (2006). "Cutting Edge: Anti-CD25 Monoclonal Antibody Injection Results in the Functional Inactivation, Not Depletion, of CD4+CD25+ T Regulatory Cells." J Immunol 176(6): 3301-3305.

Kraft, S., Novak, N. (2006). "Fc receptors as determinants of allergic reactions." Trends Immunol 27(2): 88-95.

Lahl, K., Loddenkemper, C, Drouin, C, Freyer, J, Arnason, J, Eberl, G, Hamann, A, Wagner, H, Huehn, J, Sparwasser, T. (2007). "Selective depletion of Foxp3+ regulatory T cells induces a scurfy-like disease." J Exp Med 204(1): 57-63.

Lan, C. C., Kao YH, Huang SM, Yu HS, Chen GS. (2004). "FK506 independently upregulates transforming growth factor beta and downregulates inducible nitric oxide synthase in cultured human keratinocytes: possible mechanisms of how tacrolimus ointment interacts with atopic skin." Br J Dermatol. 151(3): 679-84.

Larche, M., C. A. Akdis, and R. Valenta. (2006). "Immunological mechanisms of allergen-specific immunotherapy." Nat. Rev. Immunol 6: 761-771.

Layton, G. T., Harris, S.J., Bland, F.A., Lee, S.R., Fearn, S., Kaleta, J., Wood, M.L., Bond, A., Ward, G. (2001). "Therapeutic effects of cysteine protease inhibition in allergic

lung inflammation: inhibition of allergen-specific T lymphocyte migration." Inflamm Res **50**(8): 400-8.
Leech, M. D., Benson, R.A., De Vries, A., Fitch, P.M., Howie, S.E. (2007). "Resolution of Der p1-induced allergic airway inflammation is dependent on CD4+CD25+Foxp3+ regulatory cells." J Immunol **179**(10): 7050-8.
Lehmann, J., J. Huehn, M. de la Rosa, F. Maszyna, U. Kretschmer, V. Krenn, M. Brunner, A. Scheffold, and A. Hamann (2002). "Expression of the integrin alpha E beta 7 identifies unique subsets of CD25+ as well as CD25- regulatory T cells. ." Proc. Natl. Acad. Sci. U S A **99**: 13031-13036.
Leonardi-Bee, J., Pritchard, D., Britton, J., the Parasites in Asthma, Collaboration (2006). "Asthma and Current Intestinal Parasite Infection: Systematic Review and Meta-Analysis." Am J Respir Crit Care Med **174**(5): 514-523.
Leung-Tack, J., C. Tavera, A. Er-Raki, M. C. Gensac, and A. Colle (1990a). "Rat cystatin C: inhibitor of granulocyte phagocytic functions." Biol. Chem **371**: 255.
Leung-Tack, J., C. Tavera, M. C. Gensac, J. Martinez, and A. Colle (1990b). "Modulation of phagocytosis-associated respiratory burst by human cystatin C: role of the N-terminal tetrapeptide Lys-Pro-Pro-Arg." Exp. Cell. Res **188**: 16.
Leung, D. Y., Bieber, T. (2003). "Atopic dermatitis." Lancet **361**(9352): 151-60.
Leung, D. Y., Boguniewicz, M., Howell, M.D., Nomura, I., Hamid, Q.A. (2004). "New insights into atopic dermatitis." J Clin Invest **113**(5): 651-7.
Leung, D. Y., Hirsch RL, Schneider L, Moody C, Takaoka R, Li SH, Meyerson LA, Mariam SG, Goldstein G, Hanifin JM. (1990). "Thymopentin therapy reduces the clinical severity of atopic dermatitis." J Allergy Clin Immunol. **85**(5): 927-33.
Lima, C., Perini, A., Garcia, M.L., Martins, M.A., Teixeira, M.M., Macedo, M.S. (2002). "Eosinophilic inflammation and airway hyper-responsiveness are profoundly inhibited by a helminth (Ascaris suum) extract in a murine model of asthma." Clin Exp Allergy **33**(11): 1659-66.
Lin, Y. L., Shieh, C.C., Wang, J.Y. (2008). "The functional insufficiency of human CD4+CD25 high T-regulatory cells in allergic asthma is subjected to TNF-alpha modulation." Allergy **63**(1): 67-74.
Liu, Y.-J. (2006). "Thymic stromal lymphopoietin: master switch for allergic inflammation." J Exp Med **203**(2): 269-273.
Livak, K. J., T. D. Schmittgen (2001). "Analysis of relative gene expression data using real-time quantitative PCR and the 2exp delta delta CT method." Methods **25**: 402–408
Loke, P., MacDonald AS, Robb A, Maizels RM, Allen JE (2000). "Alternatively activated macrophages induced by nematode infection inhibit proliferation via cell-to-cell contact." Eur J Immunol **30**(9): 2669-78.
Lucius, R., Loos-Frank B. (2008). Biologie von Parasiten. Berlin, Springer Lehrbuch.
Luster, A. D. and A. M. Tager (2004). "T-cell trafficking in asthma: lipid mediators grease the way." Nat Rev Immunol **4**(9): 711-724.
Lustigman, S., B. Brotman, T. Huima, A. M. Prince, and J. H. McKerrow (1992). "Molecular cloning and characterization of onchocystatin, a cysteine protease inhibitor of Onchocerca volvulus." J. Biol. Chem **267**: 17339.
Lutz, M. B., Schuler, G. (2002). "Immature, semi-mature and fully mature dendritic cells: which signals induce tolerance or immunity?" Trends Immunol. **23**(9): 445-9.
Lynch, N., I . Hagel , M . Perez , M . Di Prisco , R . Lopez , N . Alvarez (1993). "Effect of anthelmintic treatment on the allergic reactivity of children in a tropical slum " J Allergy Clin Immunol **92**(3): 404-11.
MacDonald, AS, Maizels R, Lawrence RA, Dransfield I, Allen JE (1998). "Requirement for in vivo production of IL-4, but not IL-10, in the induction of proliferative suppression by filarial parasites." J Immunol (160): 4124–4132.

Mahnke, K., Schmitt, E., Bonifaz, L., Enk, A.H., Jonuleit, H. (2002). "Immature, but not inactive: the tolerogenic function of immature dendritic cells." Immunol Cell Biol 80(5): 477-83.

Maizels, R. M. (2005). "Infections and allergy - helminths, hygiene and host immune regulation." Curr Opin Immunol 17(6): 656-61.

Maizels, R. M., A. Balic, N. Gomez-Escobar, M. Nair, MD. Taylor, JE. Allen (2004). "Helminth parasites--masters of regulation." Immunol Rev 201: 89-116.

Maizels, R. M., D. A. P. Bundy, Selkirk ME, Smith DF, Anderson, RM (1993). "Immunological modulation and evasion by helminth parasites in human populations." Nature 365(6449): 797-805.

Maizels, R. M. and M. Yazdanbakhsh (2003). "Immune Regulation by helminth parasites: cellular and molecular mechanisms." Nat Rev Immunol 3(9): 733-744.

Mangan, N., van Rooijen, N., McKenzie, A.N. J., Fallon, P.G. (2006). "Helminth-Modified Pulmonary Immune Response Protects Mice from Allergen-Induced Airway Hyperresponsiveness." Int. Immunology 176(1): 138-147.

Mangan, N. E., Fallon, R.E., Smith, P., van Rooijen, N., McKenzie, A.N., Fallon, P.G. (2004). "Helminth infection protects mice from anaphylaxis via IL-10-producing B cells. ." J Immunol 173(10): 6346-56.

Mantovani, A., Sica, A., Locati, M. (2005). "Macrophage polarization comes of age." Immunity 23(4): 344-6.

Mantovani, A., Sozzani, S., Locati, M., Allavena, P., Sica, A. (2002). "Macrophage polarization: tumor-associated macrophages as a paradigm for polarized M2 mononuclear phagocytes " Trends Immunol 23(11): 549-55.

Marshall, F. A., Grierson, A.M., Garside, P., Harnett, W., Harnett, M. M. (2005). "ES-62, an Immunomodulator Secreted by Filarial Nematodes, Suppresses Clonal Expansion and Modifies Effector Function of Heterologous Antigen-Specific T Cells In Vivo." J Immunol 175(9): 5817-5826.

Martin, P., Leibovich, S.J. (2005). "Inflammatory cells during wound repair: the good, the bad and the ugly." Trends Cell Biol 15(11): 599-607.

Martinez, F. O., Sica, A., Mantovani, A., Locati, M. (2008). "Macrophage activation and polarization." Front Biosci 13: 453-61.

Matsuda, H., Watanabe N, Geba GP, Sperl J, Tsudzuki M, Hiroi J, Matsumoto M, Ushio H, Saito S, Askenase PW, Ra C. (1997). "Development of atopic dermatitis-like skin lesion with IgE hyperproduction in NC/Nga mice." Int Immunol. 9(3): 461-6.

Mauri, C. and M. R. Ehrenstein (2008). "The 'short' history of regulatory B cells." Trends Immunol 29(1): 34-40.

McInnes, I. B., Leung, B.P., Harnett, M., Gracie, J. A., Liew, F. Y., Harnett, W. (2003). "A Novel Therapeutic Approach Targeting Articular Inflammation Using the Filarial Nematode-Derived Phosphorylcholine-Containing Glycoprotein ES-62." J Immunol 171(4): 2127-2133.

McKee, A. S., and E. J. Pearce (2004). "CD25+CD4+ cells contribute to Th2 polarization during helminth infection by suppressing Th1 response development." J Immunol(173): 1224-1231.

Mege, J. L., Meghari, S., Honstettre, A., Capo, C., Raoult, D. (2006). "The two faces of interleukin 10 in human infectious diseases." Lancet Infect Dis. 6(9): 557-69.

Melendez, A., Harnett, MM., Pushparaj, PN., Wong, W. S. F., Tay, HK, McSharry, CP., Harnett, W. (2007). "Inhibition of Fc-epsilon-RI-mediated mast cell responses by ES-62, a product of parasitic filarial nematodes." Nat Med 13(11): 1375-1381.

Melendez, A. J. and F. B. M. Ibrahim (2004). "Antisense Knockdown of Sphingosine Kinase 1 in Human Macrophages Inhibits C5a Receptor-Dependent Signal Transduction,

Ca2+ Signals, Enzyme Release, Cytokine Production, and Chemotaxis." J Immunol **173**(3): 1596-1603.

Metwali, A., T. Setiawan, A. M. Blum, J. Urban, D. E. Elliott, L. Hang, and J. V. Weinstock (2006). "Induction of CD8+ regulatory T cells in the intestine by Heligmosomoides polygyrus infection." Am. J. Physiol. Gastrointest. Liver Physiol **291**(G): 253-259.

Miles, S., Conrad, SM., Alves, RG., Jeronimo, SMB., Mosser, DM. (2005). "A role for IgG immune complexes during infection with the intracellular pathogen Leishmania." J Exp Med **201**(5): 747-754.

Miyara, M., and S. Sakaguchi (2007). "Natural regulatory T cells: mechanisms of suppression." Trends Mol. Med **13**: 108–116.

Moncayo, A.-L. and P. J. Cooper (2006). "Geohelminth infections: Impact on allergic diseases." The International Journal of Biochemistry & Cell Biology **38**(7): 1031-1035.

Moore, K. W., R. de Waal Malefyt, R. L. Coffman, and A. O'Garra (2001). "Interleukin-10 and the interleukin-10 receptor." Annu. Rev. Immunol. **19**: 683-765.

Munn, D., Shafizadeh, E, Attwood, JT., Bondarev, I, Pashine, A, Mellor, AL. (1999). "Inhibition of T Cell Proliferation by Macrophage Tryptophan Catabolism." J Exp Med **189**(9): 1363-1372.

Munn, D. H., Pressey, J., Beall, A. C., Hudes, R., Alderson, M. R. (1996). "Selective activation-induced apoptosis of peripheral T cells imposed by macrophages. A potential mechanism of antigen-specific peripheral lymphocyte deletion." J Immunol **156**(2): 523-532.

Mutapi, F., T. Mduluza, and A. W. Roddam (2005). "Cluster analysis of schistosome-specific antibody responses partitions the population into distinct epidemiological groups." Immunol. Lett.(96): 231-240.

Nacher, M., Gay F, Singhasivanon P, Krudsood S, Treeprasertsuk S, Mazier D, Vouldoukis I, Looareesuwan S (2000). "Ascaris lumbricoides infection is associated with protection from cerebral malaria." Parasite Immunol. **22**(3): 107-13.

Nacher, M., P. Singhasivanon, et al. (2001). "Helminth infections are associated with protection from malaria-related acute renal failure and jaundice in Thailand." Am J Trop Med Hyg **65**(6): 834-836.

Nagler-Anderson, C. (2006). "Helminth-induced immunoregulation of an allergic response to food " Chem. Immunol. Allergy **90**: 1-13.

Nair, M., Gallagher, IJ., Taylor, MD., Loke, P., Coulson, PS., Wilson, R. A., Maizels, RM., Allen, JE. (2005). "Chitinase and Fizz Family Members Are a Generalized Feature of Nematode Infection with Selective Upregulation of Ym1 and Fizz1 by Antigen-Presenting Cells." Infect Immun **73**(1): 385-394.

Nakagawa, T. Y., W. H. Brissette, P. D. Lira, R. J. Griffiths, N. Petrushova, J. Stock, J. D. McNeish, S. E. Eastman, E. D. Howard, S. R. Clarke (1999). "Impaired invariant chain degradation and antigen presentation and diminished collagen-induced arthritis in cathepsin S null mice." Immunity **10**: 207.

Nakagome, K., Dohi, M., Okunishi, K., Komagata, Y., Nagatani, K., Tanaka, R., Miyazaki, J., Yamamoto, K. (2005). "In vivo IL-10 gene delivery suppresses airway eosinophilia and hyperreactivity by down-regulating APC functions and migration without impairing the antigen-specific systemic immune response in a mouse model of allergic airway inflammation." J Immunol **174**(11): 6955-66.

Negrao-Correa, D., Silveira, M.R., Borges, C.M., Souza, D.G., Teixeira, M.M. (2003). "Changes in pulmonary function and parasite burden in rats infected with Strongyloides venezuelensis concomitant with induction of allergic airway inflammation." Infect Immun **71**(5): 2607-2614.

Nicklin, M., Barrett AJ (1984). "Inhibition of cysteine proteinases and dipeptidyl peptidase I by egg-white cystatin." Biochem J 223(1): 245-53.

Niu, N., Le Goff, MK, Li, F, Rahman, M, Homer, RJ, Cohn, L. (2007). "A Novel Pathway That Regulates Inflammatory Disease in the Respiratory Tract." J Immunol 178(6): 3846-3855.

O'Garra, A., Vieira, P. (2007). "T(H)1 cells control themselves by producing interleukin-10." Nat Rev Immunol 7(6): 425-8.

Obihara, C., Beyers N, Gie RP, Hoekstra MO, Fincham JE, Marais BJ, Lombard CJ, Dini LA, Kimpen JL (2006). "Respiratory atopic disease, Ascaris-immunoglobulin E and tuberculin testing in urban South African children." Clin Exp Allergy 36(5): 640-8.

Ohmori, Y., Hamilton TA (1997). "IL-4-induced STAT6 suppresses IFN-gamma-stimulated STAT1-dependent transcription in mouse macrophages." J Immunol 159(11): 5474-82.

Ostroukhova, M., C. Seguin-Devaux, T. B. Oriss, B. Dixon-McCarthy, L. Yang, B. T. Ameredes, T. E. Corcoran, and A. Ray (2004). "Tolerance induced by inhaled antigen involves CD4+ T cells expressing membrane-bound TGF-ß and FOXP3." J. Clin. Invest(114): 28–38.

Ou, L. S., Goleva E, Hall C, Leung DY (2004). "T regulatory cells in atopic dermatitis and subversion of their activity by superantigens." J Allergy Clin Immunol 113(4): 756-63.

Ouaissi, M. A., Auriault, C., Santoro, F., Capron, A. (1981). "Interaction between Schistosoma mansoni and the complement system: role of IgG Fc peptides in the activation of the classical pathway by schistosomula." J Immunol 127(4): 1556-1559.

Owyang, A. M., Zaph C, Wilson EH, Guild KJ, McClanahan T, Miller HR, Cua DJ, Goldschmidt M, Hunter CA, Kastelein RA, Artis D. (2006). "Interleukin 25 regulates type 2 cytokine-dependent immunity and limits chronic inflammation in the gastrointestinal tract." J Exp Med 203(4): 843-9.

Ozdemir, C., Sel S, Schöll I, Yildirim AO, Bluemer N, Garn H, Ackermann U, Wegmann M, Barlan IB, Renz H, Sel S. (2007). "CD4+ T cells from mice with intestinal immediate-type hypersensitivity induce airway hyperreactivity." Clin Exp Allergy 37(10): 1419-26.

Pan, G., French, D., Mao, W., Maruoka, M., Risser, P., Lee, J., Foster, J., Aggarwal, S., Nicholes, K., Guillet, S., Schow, P., Gurney, A. L. (2001). "Forced Expression of Murine IL-17E Induces Growth Retardation, Jaundice, a Th2-Biased Response, and Multiorgan Inflammation in Mice." J Immunol 167(11): 6559-6567.

Pearce, N., Sunyer, J., Cheng, S., Chinn, S., Björkstén, B., Burr, M., Keil, U., Anderson, H.R., Burney, P. (2000). "Comparison of asthma prevalence in the ISAAC and the ECRHS. ISAAC Steering Committee and the European Community Respiratory Health Survey. International Study of Asthma and Allergies in Childhood." Eur Respir J 16(3): 420-6.

Pesce, J., Kaviratne M, Ramalingam TR, Thompson RW, Urban JF Jr, Cheever AW, Young DA, Collins M, Grusby MJ, Wynn TA (2006). "The IL-21 receptor augments Th2 effector function and alternative macrophage activation " J Clin Invest. 116(7): 2044-55.

Pfaff, A. W., Schulz-Key, H., Soboslay, P.T., Taylor, D.W., MacLennan, K., Hoffmann, W.H. (2002). "Litomosoides sigmodontis cystatin acts as an immunomodulator during experimental filariasis." Int J Parasitol 32(2): 171-8.

Pierre, P., and I. Mellman (1998). "Developmental regulation of invariant chain proteolysis controls MHC class II trafficking in mouse dendritic cells." Cell 93: 1135.

Plaisier, A. P., G. J. van Oortmarssen, J. Remme, and J. D. Habbema (1991). "The reproductive lifespan of Onchocerca volvulus on West African savanna." Acta Trop(48): 271.

Poulsen, L. K., Hummelshoj, L. (2007). "Triggers of IgE class switching and allergy development." Ann Med. **39**(6): 440 - 456.

Powrie, F., Carlino, J., Leach, M.W., Mauze, S., Coffman, R.L. (1996). "A critical role for transforming growth factor–b but not interleukin 4 in the suppression of T helper type 1–mediated colitis by CD45RB (low) CD4+ T cells." J Exp Med **183**: 2669–74.

Prussin, C., Metcalfe, D.D. (2006). "5. IgE, mast cells, basophils, and eosinophils." J Allergy Clin Immunol **117**(2 Suppl Mini-Primer): S450-6.

Rausch, S., Huehn, J., Kirchhoff, D., Rzepecka, J., Schnoeller, C., Pillai, S., Loddenkemper, C., Scheffold, A., Hamann, A., Lucius, R., Hartmann, S. (2008). "Functional analysis of effector and regulatory T cells in a parasitic nematode infection." Infect Immun **76**(5): 1908-19.

Reed, C. E. and H. Kita (2004). "The role of protease activation of inflammation in allergic respiratory diseases." J Allergy Clin Immunol **114**(5): 997-1008; quiz 1009.

Renauld, J. C., Kermouni, A., Vink, A., Louahed, J., Van Snick, J. (1995). "Interleukin-9 and its receptor: involvement in mast cell differentiation and T cell oncogenesis." J Leukoc Biol **57**(3): 353-360.

Reyes, J. L. and L. I. Terrazas (2007). "The divergent roles of alternatively activated macrophages in helminthic infections." Parasite Immunol **29**(12): 609-619.

Riese, R. J., P. R. Wolf, D. Bromme, L. R. Natkin, J. A. Villadangos, H. L. Ploegh, and H. A. Chapman (1996). "Essential role for cathepsin S in MHC class II-associated invariant chain processing and peptide loading." Immunity **4**: 357.

Romagnani, S. (2004). " Immunologic influences on allergy and the TH1/TH2 balance." J. Allergy Clin. Immunol **113**: 395-400.

Rothenberg, M. E., and S. P. Hogan (2006). "The eosinophil." Annu Rev Immunol. **24**: 147–174.

Royer, B., Varadaradjalou, S., Saas, P., Guillosson, J.J., Kantelip, J.P., Arock, M. (2001). "Inhibition of IgE-induced activation of human mast cells by IL-10." Clin Exp Allergy **31**(5): 694-704.

Rückerl, D., Hessmann, M., Yoshimoto, T., Ehlers, S., Hölscher, C. (2006). "Alternatively activated macrophages express the IL-27 receptor alpha chain WSX-1." Immunobiology **211**(6-8): 427-36.

Sakamoto, T., Miyazaki, E., Aramaki, Y., Arima, H., Takahashi, M., Kato, Y., Koga, M., Tsuchiya, S. (2004). "Improvement of dermatitis by iontophoretically delivered antisense oligonucleotides for interleukin-10 in NC//Nga mice." Gene Ther **11**(3): 317-324.

Satoguina, J., Mempel, M., Larbi, J., Badusche, M., Löliger, C., Adjei, O., Gachelin, G., Fleischer, B., Hoerauf, A. (2002). "Antigen-specific T regulatory-1 cells are associated with immunosuppression in a chronic helminth infection (onchocerciasis)." Microbes Infect **4**(13): 1291-300.

Schierack, P., Lucius, R., Sonnenburg, B., Schilling, K., Hartmann, S. (2003). "Parasite-Specific Immunomodulatory Functions of Filarial Cystatin." Infect Immun **71**(5): 2422-2429.

Schnoeller, C., Rausch, S., Pillai, S., Avagyan, A., Wittig, B. M., Loddenkemper, C., Hamann, A., Hamelmann, E., Lucius, R., Hartmann, S. (2008). "A helminth immunomodulator reduces allergic and inflammatory responses by induction of IL-10-producing macrophages." J Immunol **180**(6): 4265-72.

Schonemeyer, A., Lucius, R, Sonnenburg, B, Brattig, N, Sabat, R, Schilling, K, Bradley, J, Hartmann, S. (2001). "Modulation of Human T Cell Responses and Macrophage Functions by Onchocystatin, a Secreted Protein of the Filarial Nematode Onchocerca volvulus." J Immunol **167**(6): 3207-3215.

Schopf, L., Luccioli, S, Bundoc, V, Justice, P, Chan, C, Wetzel, BJ, Norris, HH, Urban, JF, Jr, Keane-Myers, A. (2005). "Differential Modulation of Allergic Eye Disease by Chronic and Acute Ascaris Infection." Invest Ophthalmol Vis Sci **46**(8): 2772-2780.

Scrivener, S., Yemaneberhan H, Zebenigus M, Tilahun D, Girma S, Ali S, McElroy P, Custovic A, Woodcock A, Pritchard D, Venn A, Britton J. (2001). "Independent effects of intestinal parasite infection and domestic allergen exposure on risk of wheeze in Ethiopia: a nested case-control study." Lancet **358**(9292): 1493-9.

Sereda, M., Hartmann S, Lucius R. (2008). "Helminths and allergy: the example of tropomyosin." Trends Parasitol **Apr 29 [Epub ahead of print]**

Shen, H. M., Pervaiz, S. (2006). "TNF receptor superfamily-induced cell death: redox-dependent execution." FASEB J. **20**(10): 1589-98.

Shi, G., Luo, H, Wan, X, Salcedo, TW, Zhang, J, Wu, J. (2002). "Mouse T cells receive costimulatory signals from LIGHT, a TNF family member." Blood **100**(9): 3279-3286.

Shpacovitch, V., Feld, M., Bunnett, N. W., Steinhoff, M. (2007). "Protease-activated receptors: novel PARtners in innate immunity." Trends Immunol **28**(12): 541-50.

Siewert, C., Menning A, Dudda J, Siegmund K, Lauer U, Floess S, Campbell DJ, Hamann A, Huehn J. (2007). "Induction of organ-selective CD4+ regulatory T cell homing." Eur J Immunol **37**(4): 978-89.

Simon, H. U. (1999). "New insights into the pathogenesis of asthma." Curr Probl Dermatol **28**: 124-8.

Simon, D., Braathen LR, Simon HU. (2007). "Increased lipopolysaccharide-induced tumour necrosis factor-alpha, interferon-gamma and interleukin-10 production in atopic dermatitis." Br J Dermatol. **157**(3): 583-6

Smith, P., Fallon, RE, Mangan, NE, Walsh, CM, Saraiva, M, Sayers, JR, McKenzie, ANJ, Alcami, A, Fallon, PG. (2005). "Schistosoma mansoni secretes a chemokine binding protein with antiinflammatory activity." J Exp Med **202**(10): 1319-1325.

Smith, P., Mangan, N. E., Walsh, C. M., Fallon, R. E., McKenzie, A. N., van Rooijen, N., Fallon, P. G. (2007). "Infection with a helminth parasite prevents experimental colitis via a macrophage-mediated mechanism." J Immunol **178**(7): 4557-66.

Smith, P., Walsh, C.M., Mangan, N.E., Fallon, R.E., Sayers, J.R., McKenzie, A.N., Fallon, P.G. (2004). "Schistosoma mansoni worms induce anergy of T cells via selective up-regulation of programmed death ligand 1 on macrophages." J Immunol **173**(2): 1240-8.

Smits, H. H., H. Hammad, M. van Nimwegen, T. Soullie, M.A. Willart, E. Lievers, J. Kadouch, M. Kool, J. Kos-van Oosterhoud, A.M. Deelder, B.N. Lambrecht, and M. Yazdanbakhsh (2007). "Protective effect of Schistosoma mansoni infection on allergic airway inflammation depends on the intensity and chronicity of infection." J. Allergy Clin. Immunol **120**: 932-940.

Sokol, J. P., and W. P. Schiemann (2004). "Cystatin C antagonizes transforming growth factor ß signaling in normal and cancer cells." Mol. Cancer Res. **2**: 183–195.

Souza, V. M., Jacysyn, J.F., Macedo, M.S. (2004). "IL-4 and IL-10 are essential for immunosuppression induced by high molecular weight proteins from Ascaris suum." Cytokine **28**(2): 92-100.

Spellberg, B., Edwards, Jr. J.E. (2001). "Type 1/Type 2 Immunity in Infectious Diseases." Clinical Infectious Diseases **32**(1): 76-102.

Spergel, J. M., Mizoguchi E, Brewer JP, Martin TR, Bhan AK, Geha RS. (1998). "Epicutaneous sensitization with protein antigen induces localized allergic dermatitis and hyperresponsiveness to methacholine after single exposure to aerosolized antigen in mice." J Clin Invest. **101**(8): 1614-22.

Stephens, L. A. and S. M. Anderton (2006). "Comment on "Cutting Edge: Anti-CD25 Monoclonal Antibody Injection Results in the Functional Inactivation, Not Depletion, of CD4+CD25+ T Regulatory Cells"." J Immunol **177**(4): 2036-.

Stock, P., DeKruyff, R.H., Umetsu, D.T. (2006). "Inhibition of the allergic response by regulatory T cells." Curr Opin Allergy Clin Immunol **6**(1): 12-6.

Stout, R. D. and J. Suttles (2004). "Functional plasticity of macrophages: reversible adaptation to changing microenvironments." J Leukoc Biol **76**(3): 509-513.

Strachan, D. (1989). "Hay fever, hygiene, and household size." BMJ **299**(6710): 1259-60.

Subramanian, S., Stolk WA, Ramaiah KD, Plaisier AP, Krishnamoorthy K, Van Oortmarssen GJ, Dominic Amalraj D, Habbema JD, Das PK (2004). "The dynamics of Wuchereria bancrofti infection: a model-based analysis of longitudinal data from Pondicherry, India." Parasitology **128**(Pt 5): 467-82.

Sumiyoshi, K., Nakao A, Ushio H, Mitsuishi K, Okumura K, Tsuboi R, Ra C, Ogawa H. (2002). "Transforming growth factor-beta1 suppresses atopic dermatitis-like skin lesions in NC/Nga mice." Clin Exp Allergy **32**(2): 309-14.

Summers, R. W., D. E. Elliot, J. F. Jr. Urban, R. A. Thompson, and J. V. Weinstock (2005a). "Trichuris suis therapy for active ulcerative colitis: a randomized controlled trial." Gastroenterology(128): 825–832.

Summers, R. W., D. E. Elliot, J. F. Jr. Urban, R. Thompson, and J. V. Weinstock (2005b). "Trichuris suis therapy in Crohn's disease." Gut (54): 87–90.

Sun, Y., Blink, SE, Liu, W, Lee, Y, Chen, B, Solway, J, Weinstock, J, Chen, L, Fu, Y. (2006). "Inhibition of Th2-Mediated Allergic Airway Inflammatory Disease by CD137 Costimulation." J Immunol **177**(2): 814-821.

Tamachi, T., Maezawa Y, Ikeda K, Kagami S, Hatano M, Seto Y, Suto A, Suzuki K, Watanabe N, Saito Y, Tokuhisa T, Iwamoto I, Nakajima H. (2006). "IL-25 enhances allergic airway inflammation by amplifying a TH2 cell-dependent pathway in mice." J Allergy Clin Immunol. **118**(3): 606-14.

Tamada, K., Shimozaki, K, Chapoval, AI, Zhai, Y, Su, J, Chen, S, Hsieh, S, Nagata, S, Ni, J, Chen, L. (2000). "LIGHT, a TNF-Like Molecule, Costimulates T Cell Proliferation and Is Required for Dendritic Cell-Mediated Allogeneic T Cell Response." J Immunol **164**(8): 4105-4110.

Taylor, A., J. Verhagen, K. Blaser, M. Akdis, and C. A. Akdis (2006). "Mechanisms of immune suppression by interleukin-10 and transforming growth factor-beta: the role of T regulatory cells." Immunology **117**: 433-442.

Taylor, M., Harris, A, Nair, MG., Maizels, RM., Allen, JE. (2006). F4/80+ Alternatively Activated Macrophages Control CD4+ T Cell Hyporesponsiveness at Sites Peripheral to Filarial Infection. **176**: 6918-6927.

Terrazas, L. I., Walsh, K. L., Piskorska, D., McGuire, E., Harn, D. A., Jr. (2001). "The Schistosome Oligosaccharide Lacto-N-neotetraose Expands Gr1+ Cells That Secrete Anti-inflammatory Cytokines and Inhibit Proliferation of Naive CD4+ Cells: A Potential Mechanism for Immune Polarization in Helminth Infections." J Immunol **167**(9): 5294-5303.

Thomas, W. R., Smith WA, Hales BJ, Mills KL, O'Brien RM. (2002). "Characterization and immunobiology of house dust mite allergens." Int Arch Allergy Immunol **129**(1): 1-18.

Toda, M., Leung, D.Y., Molet, S., Boguniewicz, M., Taha, R., Christodoulopoulos, P., Fukuda, T., Elias, J.A., Hamid, Q.A. (2003). "Polarized in vivo expression of IL-11 and IL-17 between acute and chronic skin lesions." J Allergy Clin Immunol **111**(4): 875-81.

Tournoy, K. G., Kips, J.C., Pauwels, R.A. (2000). "Endogenous interleukin-10 suppresses allergen-induced airway inflammation and nonspecific airway responsiveness." Clin Exp Allergy **30**(6): 775-83.

Trautmann, A., Akdis, M., Klunker, S., Blaser, K., Akdis, C. A. (2001). "Role of Apoptosis in Atopic Dermatitis." International Archives of Allergy and Immunology **124**(1-3): 230-232.

Trivedi, S. G., Lloyd, C. M. (2007). "Eosinophils in the pathogenesis of allergic airways disease." Cell Mol Life Sci **64**(10): 1269-89.

Trujillo-Vargas, C. M., M. Werner-Klein, G. Wohlleben, T. Polte, G. Hansen, S. Ehlers, and K. J. Erb (2007). "Helminth-derived Products Inhibit the Development of Allergic Responses in Mice." Am. J. Respir. Crit. Care Med **175**: 336-344.

Umetsu, D. T., DeKruyff, R.H. (2006a). "Immune dysregulation in asthma." Curr Opin Immunol **18**(6): 727-32.

Umetsu, D. T., DeKruyff, R.H. (2006b). "The regulation of allergy and asthma." Immunol Rev **212**: 238-55.

Umetsu, D. T., McIntire, J.J., Akbari, O., Macaubas, C., DeKruyff, R.H. (2002). "Asthma: an epidemic of dysregulated immunity." Nat Immunol **3**(8): 715-20.

Urban, B. C., N. Willcox, and D. J. Roberts (2001). "A role for CD36 in the regulation of dendritic cell function." Proc. Natl. Acad. Sci. USA **98**: 8750–8755.

Urry, Z., E. Xystrakis, and C. M. Hawrylowicz (2006). "Interleukin-10-secreting regulatory T cells in allergy and asthma." Curr. Allergy Asthma Rep **6**: 363-371.

van Beelen, A. J., Teunissen, M.B., Kapsenberg, M.L., de Jong, E.C. (2007). "Interleukin-17 in inflammatory skin disorders." Curr Opin Allergy Clin Immunol **7**(5): 374-81.

van den Biggelaar, A., Rodrigues, LC., van Ree, R., van der Zee, JS., Hoeksma-Kruize, YCM., Souverijn, JHM., Missinou, MA., Borrmann, S., Kremsner, PG., Yazdanbakhsh, M. (2004). "Long-Term Treatment of Intestinal Helminths Increases Mite Skin-Test Reactivity in Gabonese Schoolchildren." J Infect Dis **189**(5): 892-900.

van den Biggelaar A., R. v. R., L . Rodrigues , B . Lell , A . Deelder , P . Kremsner , M . Yazdanbakhsh (2000). "Decreased atopy in children infected with a role for parasite-induced interleukin-10." Lancet **356** (9243): 1723 - 1727.

van den Biggelaar, A. H. J., Lopuhaa, C., van Ree, R.,van der Zee, J. S., Jans, J., Hoek, A., Migombet, B., Borrmann, S., Luckner, D., Kremsner, P. G., Yazdanbakhsh, M. (2001). "The Prevalence of Parasite Infestation and House Dust Mite Sensitization in Gabonese Schoolchildren." International Archives of Allergy and Immunology **126**(3): 231-238.

van der Kleij, D., Latz, E, Brouwers, JFHM, Kruize, YCM, Schmitz, M, Kurt-Jones, EA, Espevik, T, de Jong, EC, Kapsenberg, ML, Golenbock, DT, Tielens, AGM, Yazdanbakhsh, M. (2002). "A novel host-parasite lipid cross-talk. Schistosomal lyso-phosphatidylserine activates toll-like receptor 2 and affects immune polarization " J Biol Chem **277**(50): 48122-48129.

van Die, I., Cummings, R.D. (2006). "Glycans modulate immune responses in helminth infections and allergy." Chem Immunol Allergy **90**: 91-112.

van Riet, E., Hartgers FC, Yazdanbakhsh M (2007). "Chronic helminth infections induce immunomodulation: consequences and mechanisms." Immunobiology **212** (6): 475-490.

Van Rooijen, N., Sanders A (1994). "Liposome mediated depletion of macrophages: mechanism of action, preparation of liposomes and applications." J Immunol Methods **174**(1-2): 83-93.

Velasco, G., M. Campo, O. J. Manrique, A. Bellou, H. He, R. S. Arestides, B. Schaub, D. L. Perkins, and P. W. Finn (2005). "Toll-like receptor 4 or 2 agonists decrease allergic inflammation." Am. J. Respir. Cell. Mol. Biol. **32**: 218–224.

Veldhoen, M., Stockinger, B. (2006). "TGFbeta1, a "Jack of all trades": the link with pro-inflammatory IL-17-producing T cells." Trends Immunol. 27(8): 358-61.

Vercelli, D., Martinez, FD. (2006). "The Faustian bargain of genetic association studies: bigger might not be better, or at least it might not be good enough." J Allergy Clin Immunol 117(6): 1303-5.

Verdot, L., G. Lalmanach, V. Vercruysse, J. Hoebeke, F. Gauthier, and B. Vray (1999). "Chicken cystatin stimulates nitric oxide release from interferon-gamma-activated mouse peritoneal macrophages via cytokine synthesis." Eur. J. Biochem 266: 1111.

Verhagen, J., Akdis M, Traidl-Hoffmann C, Schmid-Grendelmeier P, Hijnen D, Knol EF, Behrendt H, Blaser K, Akdis CA. (2006). "Absence of T-regulatory cell expression and function in atopic dermatitis skin." J Allergy Clin Immunol 117(1): 176-83.

Vignola, A., Chanez, P, Chiappara, G, Merendino, A, Pace, E, Rizzo, A, la Rocca, AM, Bellia, V, Bonsignore, G, Bousquet, J. (1997). "Transforming Growth Factor-beta Expression in Mucosal Biopsies in Asthma and Chronic Bronchitis." Am J Respir Crit Care Med 156(2): 591-599.

Wan, X., Zhang, J, Luo, H, Shi, G, Kapnik, E, Kim, S, Kanakaraj, P, Wu, J. (2002). "A TNF Family Member LIGHT Transduces Costimulatory Signals into Human T Cells." J Immunol 169(12): 6813-6821.

Wang, C., Nolan TJ, Schad GA, Abraham D (2001). "Infection of mice with the helminth Strongyloides stercoralis suppresses pulmonary allergic responses to ovalbumin." Clin Exp Allergy 31(3): 495-503.

Wang, G., Savinko, T., Wolff, H., Dieu-Nosjean, M. C., Kemeny, L., Homey, B., Lauerma, A. I., Alenius, H. (2007). "Repeated epicutaneous exposures to ovalbumin progressively induce atopic dermatitis-like skin lesions in mice." Clin Exp Allergy 37(1): 151-161.

Warfel, A., Zucker-Franklin D, Frangione B, Ghiso J. (1987). "Constitutive secretion of cystatin C (gamma-trace) by monocytes and macrophages and its downregulation after stimulation." J Exp Med 166(6): 1912-7.

Watts, C. (2001). "Antigen processing in the endocytic compartment." Curr Opin Immunol 13(1): 26-31.

Weaver, C. T., Hatton RD, Mangan PR, Harrington LE. (2007). "IL-17 family cytokines and the expanding diversity of effector T cell lineages." Annu Rev Immunol. 25: 821-52.

Weinstock, J. V. (2006). "Helminths and Mucosal Immune Modulation." Ann N Y Acad Sci 1072(1): 356-364.

Whelan, M., Harnett, M. M., Houston, K.M., Patel, V., Harnett, W., Rigley, K. P. (2000). "A Filarial Nematode-Secreted Product Signals Dendritic Cells to Acquire a Phenotype That Drives Development of Th2 Cells." J Immunol 164(12): 6453-6460.

Wills-Karp, M., Santeliz, J., Karp, CL. (2001). "The germless theory of allergic disease: revisiting the hygiene hypothesis." Nat Rev Immunol 1(1): 69-75.

Wilson, E. H., Katz, E., Goodridge, H.S., Harnett, M.M., Harnett, W. (2003). "In vivo activation of murine peritoneal B1 cells by the filarial nematode phosphorylcholine-containing glycoprotein ES-62." Parasite Immunol 25(8-9): 463-6.

Wilson, M., Taylor, MD., Balic, A., Finney, C.A. M., Lamb, J.R., Maizels, RM. (2005). "Suppression of allergic airway inflammation by helminth-induced regulatory T cells." J Exp Med 202(9): 1199-1212.

Wohlleben, G., Trujillo C, Müller J, Ritze Y, Grunewald S, Tatsch U, Erb KJ (2004). "Helminth infection modulates the development of allergen-induced airway Inflammation." Int. Immunology 16(4): 585-596.

Wördemann, M., Diaz RJ, Heredia LM, Collado Madurga AM, Ruiz Espinosa A, Prado RC, Millan IA, Escobedo A, Rojas Rivero L, Gryseels B, Gorbea MB, Polman K. (2008).

"Association of atopy, asthma, allergic rhinoconjunctivitis, atopic dermatitis and intestinal helminth infections in Cuban children." <u>Trop Med Int Health.</u> **13**(2): 180-6.

Wu, W., Mosteller, RD, Broek, D. (2004). "Sphingosine Kinase Protects Lipopolysaccharide-Activated Macrophages from Apoptosis." <u>Mol Cell Biol</u> **24**(17): 7359-7369.

Yang, J., Castle, BE, Hanidu, A, Stevens, L, Yu, Y, Li, X, Stearns, C, Papov, V, Rajotte, D, Li, J. (2005). "Sphingosine Kinase 1 Is a Negative Regulator of CD4+ Th1 Cells." <u>J Immunol</u> **175**(10): 6580-6588.

Yazdanbakhsh, M., P. G. Kremsner, et al. (2002). "Allergy, Parasites, and the Hygiene Hypothesis." <u>Science</u> **296**(5567): 490-494.

Zaccone, P., Burton, O.T., Cooke, A. (2008). "Interplay of parasite-driven immune responses and autoimmunity." <u>Trends Parasitol</u> **24**(1): 35-42.

Zelenay, S. and J. Demengeot (2006). "Comment on "Cutting Edge: Anti-CD25 Monoclonal Antibody Injection Results in the Functional Inactivation, Not Depletion, of CD4+CD25+ T Regulatory Cells"." <u>J Immunol</u> **177**(4): 2036-a-2037.

Zhang, X., D. M. Mosser (2008). "Macrophage activation by endogenous danger signals." <u>J Pathol</u> **214**(2): 161-178.

Die VDM Verlagsservicegesellschaft sucht für wissenschaftliche Verlage abgeschlossene und herausragende

Dissertationen, Habilitationen, Diplomarbeiten, Master Theses, Magisterarbeiten usw.

für die kostenlose Publikation als Fachbuch.

Sie verfügen über eine Arbeit, die hohen inhaltlichen und formalen Ansprüchen genügt, und haben Interesse an einer honorarvergüteten Publikation?

Dann senden Sie bitte erste Informationen über sich und Ihre Arbeit per Email an *info@vdm-vsg.de*.

Sie erhalten kurzfristig unser Feedback!

VDM Verlagsservicegesellschaft mbH
Dudweiler Landstr. 99　　　　　　　Telefon　+49 681 3720 174
D - 66123 Saarbrücken　　　　　　Fax　　　+49 681 3720 1749
www.vdm-vsg.de

Die VDM Verlagsservicegesellschaft mbH vertritt

Printed by Books on Demand GmbH, Norderstedt / Germany